The A Chaos Magic Paradigm

By Peter J Carroll "Stokastikos"[0]

Past Grandmaster of the
Magical Pact of the Illuminates of Thanateros.[1]

Chancellor of Arcanorium College.[2]

Copyright © 2008 Mandrake & Peter J Carroll

First edition 08/08/08

All rights reserved. No part of this work may be reproduced or utilized in any form by any means electronic or mechanical, including *xerography, photocopying, microfilm,* and *recording,* or by any information storage system without permission in writing from the publishers.

Contents

Acknowledgements --------------------------------------- 5
1. Apophenia - Introduction ----------------------------- 7
2. Panpsychism - Philosophy ---------------------------- 11
3. Multimind - Psychology ------------------------------ 28
4. Neopantheism - DIY Religion ------------------------- 46
5. Metadynamics - Practical Magic ---------------------- 64
6. Non-Singularity - Cosmology ------------------------- 88
7. Illumination? -------------------------------------- 104
8. An Invocation of Apophenia ------------------------- 114

Appendix I. Three-dimensional time and quantum geometry. -- 127
Appendix II. Hypersphere from Radius Excess ---------- 135
Appendix III. The Hyperspherical Universe. ----------- 138
Appendix IV. The Shape of the Universe --------------- 145
Appendix V. Apophenia's Birthday --------------------- 148

Epilogue --- 153
Notes, References, and Bibliography ------------------ 155
Index -- 157

Invoking Sigil of Apophenia. By the authors hand, - after the method of Ausin Osman Spare.

Acknowledgements

The author wishes to thank the Muse in all her guises.

Ingrid Glaw for her excellent interpretations of the chapter themes which appear in her illustrations; and David Gough for so generously allowing us to use an image of his painting 'Gods and Monsters' for the cover. These artists can be contacted at:
http://iggygirl.deviantart.com
http://www.davidgoughart.com

The Maybelogic Academy (www.maybelogic.org)and my Psychonaut and Independence crews who sailed the ether with me online, they inspired me to write again. Arcanorium College (www.arcanoriumcollege.com) my thanks to staff and members, particularly 8 Wasps and Res, who read the drafts, revised my idiosyncratic spellings and drew my attention to areas requiring clarification. Rarely since the Bible has a book had so many people arguing about its contents.

A number of world class physicists who replied to enquiries about the hypotheses in this book during their construction. Most of them receive a daily avalanche of nut mail; nevertheless some gave helpful criticism or further references to chase up. Some expressed concern at the high degree of symmetry in the hypothesis, most thought it needed more maths to fully justify it, but none could see any obvious holes in it despite that it appeared a bit of a longshot. In return for their kindness I shall not cast a shadow over anyone's professional eminence by mentioning any names.

Mogg Morgan of Mandrake for his editorial advice.

WARNING.

This book of Magic also contains a certain amount of Physics.

"A Witch is a Rebel in Physics"
Thomas Vaughan, Anthroposophia Theomagica, 1650.

Chapter 1
Apophenia - Introduction

Physics means no more than a set of ideas about how the world works; everybody has some sort of theory of physics, based on anything from simple experience and intuition to sophisticated experiment and hypothesis.

As magic works, at least occasionally, it must form part of any complete theory of how the world works.

I regard physics as that subset of magic that works fairly reliably. I regard magic, in the traditional sense, as a kind of physics that we strive to understand and render more reliable. So it all comes down to the same thing, a quest to understand and manipulate the world with a self-consistent and coherent theory.

Magic implies an extension of 'ordinary' physics which should tell us more about how the universe works and perhaps suggest how we can refine the theory and practise of magic itself.

As the third millennium begins, most of the certainties that have guided thought for the previous two millennia now begin to look very questionable. A revolution started to germinate in the 20th century with the advent of Relativity and Quantum physics and the birth of a completely new esoteric theory, Chaoism.

This book advances the thesis that all three of those new fields now converge to smash most of the assumptions that have guided humanity for centuries.

Welcome to the paradigm crash of the third millennium.

Magic and Science stand poised to overturn just about everything we believed about life, reality mind, consciousness, religion, causality, and the universe. If the word 'Magic' sounds too outrageous, then substitute psychological and parapsychological technology instead.

Of course for the 93% of humanity that eschews abstract thought, the paradigm shift will come slowly, as the new insights filter down from those Illuminati who use them to practical effect.

Each of the following chapters of this book begins with the assassination of an idea that has held for decades, centuries or millennia. Each chapter then seeks Apophenia in an alternative to the demolished idea.

Apophenia means finding pattern or meaning where others don't. Feelings of revelation and ecstasis usually accompany it. It has some negative connotations in psychological terminology when it implies finding meaning or pattern where none exists; and some positive ones when it implies finding something important, useful, or beautiful. It thus links creativity and psychosis, genius and madness.

A talent for Apophenia frequently characterises magicians, mystics and occultists. At its best it opens up whole new fields of human endeavour, it has close associations with Pareidolia, the mistaking of pieces of rope for snakes, seeing goats, bulls, and virgins in the positions of stars and in the personalities of people, the construction of unreasonable conspiracy theories, and the theology of sky fairies. Nevertheless Pareidolia plays its part in the development of art and religion.

By convention we tend to regard inspiration as female because of its association with holistic right cerebral hemisphere brain activity, rather than with left hemisphere linear thought.

Apophenia does not always come when we call her, sometimes she rejects our seductions and entreaties, sometimes she calls when we're out, (of our heads), sometimes not. Sometimes her mad sister Pareidolia comes instead.

Chaoism seeks to explore the inner riches and to expand the Inner Mythos, the pantheon of powers within. For decades I pursued the mythos of Ouranos, the magician identity that lay beyond the soap-opera of the seven classical motivations of sex-death, fear-desire, love-war, and ego. Lately I have come to realise that I love Apophenia, the female aspect of the Ouranian current, above all else.

(Uranus-Ouranos lies outside of the classical seven planets and their fancifully attributed motivations, and thus provides a useful counterpoint to the 'normal' solar identity or ego).

I have a modest taste in deities. I reject the hyper-inflated ego model of any monotheistic deity with a big 'D'.

Some people believe that someone created a universe with a volume of at least a trillion-trillion cubic light years, containing at least a billion stars for every human, set in a radiation blasted vacuum. They furthermore believe that this 'person' gets either pleased or angry with them personally if they eat pork on a Friday, or masturbate on a Sunday, or massacre the enemies of the faith on a Wednesday, or whatever their current infallible theology dictates. This sounds like serious mental illness, a kind of megalomania by proxy.

I prefer household gods, the ones that I can find inside my own head, and sometimes inside other people's heads as well.

Above all I have come to love Apophenia, the goddess who showed me how to find meaning in the last place that I expected to find it, in a universe which runs on the only truly fair and equitable system, pure chance, randomness and chaos.

I would kill for her, in fact I have attempted murder many times in her honour. See the following chapters. Being, Self, God, Causality, and Singularity; all of them get flayed upon her altar to see what illuminations and magical possibilities lie beyond.

Stokastikos,

Peter J Carroll. Albion Southwest. 2008.

Chapter 2
Panpsychism - Philosophy

This chapter begins with a deconstruction and demolition of the concept of 'Being' and proceeds through an examination of Pantheism to seek an Apophenia in the paradigm of Quantum Panpsychism and its use in Magic.

Part 1. The Metaphysics of Non-Being

Metaphysics means the set of assumptions underlying the way we interpret the phenomena that we perceive. Big assumptions like the existence of mind, matter, gods, causality, and randomness all fall into this category.

The word phenomena (or phenomenon for singular), merely denotes events that we perceive. By refraining from talking about the 'things' we perceive we avoid making too many initial assumptions, in particular we avoid the questionable concept of 'thing-ness'.

Can we find 'The universe in a grain of sand'?

Well perhaps, but a stone seems easier to visualise.

Cursory examinations of simple phenomena like stones, suggest that on their own, they don't actually do anything much.

From such simple observations we have built entirely false models of reality with languages and philosophies to match.

A more detailed examination of a stone requires devising artificial extensions to our rather meagre sensory capabilities. For a few hundred thousand years we got used to the idea of stones not really doing anything much on their own, but in the

The Eagle Dragon of the Primal Chaos*
Prometheus-Lucifer,
Challenging the heavens
with the fires of the Titans.
* Mass of Chaos, *Liber Null*

last century or so we have come to realise that even the simplest piece of stone does a great deal.

Beneath the hard apparently immobile exterior of any piece of stone lies a swirling world of high energy activity conducting itself at astonishing speeds.

For a start, a stone actively interacts with light, selectively absorbing some frequencies and emitting others, which means that it exhibits a distinctive colour. The molecules within the stone vibrate at a rate dependant on its temperature. If they ceased to vibrate, its temperature would drop to absolute zero and it would shrink towards zero size. The electrons within the atoms that make up the molecules of the stone have very high orbital velocities, of the order of hundreds of miles or kilometres per hour, and they also undergo a complicated sort of spin as they orbit. In the nuclei of the atoms of the stone very complicated processes involving even higher energies proceed ceaselessly. The stone also interacts with the whole universe gravitationally, fractionally bending space and time around itself and responding to the spacetime curvature of bigger objects like planets and stars.

So all in all, a stone consists of many processes. If you push it, it pushes back with its inertia, if you try to poke it, its electrons move to repel the ones in your finger.

We cannot really ask what a stone 'is', we can only ask what it does, or what it resembles, or how we feel about it.

We have no reason to suppose that it consists of anything other than the totality of what it does.

However our meagre unaided sensory capabilities encourage our simpler brain programs to conceptualise a stone as having some sort of static state of 'being' because we cannot directly perceive, or easily conceive of, most of the doing going on. This misconception of 'being' leads to the erection of entirely fallacious philosophies and assumptions. These have serious practical consequences, and they have killed millions of people. (Wait a few pages to find out how).

Popular science authors seem to delight in revealing that the atoms, which make up the world and us and the stars, consist almost entirely of empty space. They often use the analogy that an atom magnified to the size of a concert hall would have a nucleus the size of a pea in the orchestra pit, with pinhead sized electrons orbiting at the distance of the rear stalls.

This rather depends on what you mean by 'empty space'. It seems unlikely that any such thing as empty space actually exists. Although electrons sometimes behave as dimensionless points, when they orbit the nuclei of atoms they behave like diffuse clouds spread right round their orbital paths. A stone also exhibits a certain amount of gravity, and gravity consists of a curvature in space and time. We do not normally notice the spacetime curvature of stones, but really big ones, stones the size of moons or planets, do exhibit an unmistakable curvature which causes smaller objects to fall towards them or to stick to their surfaces. This curvature extends as far as the universe extends, so in one sense, any object stretches right across the universe. The apparent limiting surface of an object arises in our perception only because of short-range electrostatic forces between electrons and because of interactions between electrons and light. Creatures that perceived only gravity would experience any object as a phenomenon that extended from its centre with gradually diminishing intensity to the limits of the universe.

The short range 'forces' inside an atom probably also consist of a special sort of spacetime curvature, and so in a sense they completely fill it up. In other words spacetime has a structure which arises from the presence of matter within it, or conversely the curvature of spacetime appears to us as the presence of matter.

The idea of subatomic particles having some kind if definite size makes little sense anyway. They have measurable wavelengths which can determine the size of hole they can go through, but wavelength tends to decrease as the mass of quantum particles or their energy or speed increases. Electrons in atoms can absorb or emit photons (light quanta) which appear to us as vastly 'larger' in some sense, than the electrons themselves.

Our unaided senses tend to encourage us to model space and time as Privative phenomena, (which merely consist of the absence of events). Death for example does not exist in a positive sense, it consists merely of the absence of life activity, and similarly Darkness consists merely of the absence of light quanta activity.

However we can no longer regard space as merely the absence of stuff, and time as merely an interval between events. Spacetime has a structure defined by the presence of matter and energy, large concentrations of matter distort spacetime by bending it, and travel at very high speeds measurably deforms it.

Thus if we want to think clearly about the universe in which we find ourselves, we should no longer regard space and time as some sort of passive stage on which objects have their 'being' and execute various actions under the influence of energy.

On close inspection, the whole 'thingness' of objects that we conceptualise on the macroscopic (human size) scale just evaporates.

No phenomenon exhibits 'being'. All phenomena consist of ongoing processes; they consist of various doings.

About two and a half thousand years ago, the early Buddhist philosophers recognised the impermanence and the illusory nature, and hence the 'emptiness' of all phenomena except change itself. From the observation that most phenomena change if you observe them for long enough, they proceeded by induction to the idea that they all do.

Less patient western thinkers simply assumed 'being' and then eventually, after frantic researches lasting centuries, to find out what 'things' actually 'are', they found that every phenomenon they examined underwent change. The universe itself changes with time. Stars explode or collapse eventually; worlds accrete from dust and gas and cannot persist forever.

Westerners frequently misinterpret the Buddhist idea of the illusory nature of reality as more or less equivalent to the denigration of the material plane in favour of the spiritual plane, which occurs in much monotheistic thought. Strict Buddhists however, regard the 'spiritual' as impermanent as the 'material'. Nevertheless, the austere core ideas of Buddhism rarely manifest in common practise and belief. Wherever you look they usually appear dressed in local custom and contaminated with superstition because people generally prefer folksy comforting religions and mysterious rituals to difficult ideas.

A stone does not have any kind of 'being' underlying what it does. It consists entirely of its doing, and if it ceased such doing, 'it' would not have any kind of existence.

Any so-called attribute of 'being' invariably arises from some kind of doing if you examine it closely enough.

We inhabit a universe of events, not a universe full of things. Phenomena can give the macroscopic impression of having 'being' or 'thingness' but only because they actually consist of ongoing processes.

I don't know about you, but I certainly do not have any sort of intrinsic being apart from what I do. In my youth I exhibited various behaviours, performed various thoughts, emotions, and acts, and expressed various opinions and ideals. In my middle years I now do different activities, my body looks different and it contains hardly any of the atoms or molecules that it did decades ago. I seem to have irretrievably lost many memories of trivial or boring events; and my mind now contains many things that it did not in my youth. When, or if, I get older, the older version may differ markedly from the current one in what it does.

Thus I conclude that I do not have any sort of 'being', I consist only of the totality of what I do. I proceed through time as a process.

The concept of 'being' may seem a harmless enough but rather sloppy and inaccurate way of modelling reality but it leads to appalling consequences. Every use of the words of the verb 'to be', like 'is' or 'are', conceals a false or questionable premise.

The statement 'Today 'is' Wednesday' has only limited applicability, it may well not apply to the situation on the other side of the planet. The assertion that 'Pete 'is' stupid' has an outrageous generality. Does he invariably exhibit stupid behaviour?

The assertion that Brown, White, Black, Yellow, Jewish, or French people 'are' dirty, clever, devious, brave, stupid, subhuman, evil, or whatever, leads to irrational thoughts and ghastly consequences, despite that some people within those groups, or indeed within any groups, may exhibit such behaviours at some times under various circumstances.

If we want to philosophise with clarity we can not say that any phenomena 'is' any other phenomena. We can only speak of actions, resemblances, and differences.

If we try and define what any phenomenon 'is' we merely apply a label to it, or say what its behaviour resembles. We can only define phenomena in terms of their resemblance to other phenomena and by implication, to what they do.

Any statement about what anything 'is' only has utility to the extent that it implies what it does.

When we speak of what any phenomenon 'does' we actually imply what we think it has done and what we think it will do.

'Being' exists only as a neurological and linguistic illusion.

The behaviour of quantum phenomena barely resembles the behaviour of anything else at all. Thus all attempts to define them in terms of what they 'are' end in failure.

At best we can hope to describe what they do on the basis of what we think they have done have done and what we expect them to do. That actually that applies to every single phenomenon in the universe if we apply strict logic.

The assumption that an electron is, or ought to be, either a wave or a particle, or indeed that it 'is' anything, renders quantum physics completely incomprehensible.

The concept of 'being' implies some kind of metaphysical essence or quality in a phenomenon which exists somewhat independently of what we actually observe it doing.

This being-doing duality leads directly to the misconception of a spirit-matter dualism which underpins nearly all religious ideas, and to a mind-matter or to a mind-body dualism which gives rise to insoluble but illusory problems and paradoxes in philosophy, psychology, and in our ideas about consciousness.

So the seemingly innocuous idea of 'being' encourages sloppy inaccurate thinking and prejudice, it allows us to create idiotic religious ideas, it prevents us from understanding how the universe works, and it renders us incomprehensible to ourselves.

Language structures thought, to at least the same degree that it reflects thought. Only with the greatest of difficulty can we formulate a thought which involves a concept for which we lack a word. Every word you do not understand represents an idea that you cannot easily have, but on the other hand, words can give a spurious reality to concepts that have no correlate in the real world at all.

In particular the subject-verb-object sentence structure of the English language, and most other languages, encourages users to think in terms of the subject having some sort of separate 'being' from what it does.

The exegesis presented in this book avoids the use of such words as 'am', 'is', and 'are', except in parenthesis for illustrative purposes. It similarly avoids the word 'was' for reasons which appear in Chapter 5.

The abandonment of the language and concept of 'being' leads to a strict Monism, which eliminates any kind of spirit-matter or mind-body dualism.

If we assert the reality of both spirit and matter, or of mind and matter we should only do so in terms of what these phenomena actually do, not what we suppose they 'are'.

When we look at what kind of events actually occur, we find that we need only a single class of phenomena to account for it, and it makes no difference whether we call it spirit or mind or matter.

Let's leave spirit out of the argument for a while because it does not seem to do anything except allegedly act as the mind of supposedly superhuman creatures.

Now that we know a lot about how the body works, we have no reason to suppose that the body consists of anything other than matter. Thus we need only consider the mind-matter duality.

Most people subjectively experience the actions of mind as quite separate from the activities of matter, although our ancestors and our childhood selves often did not make such rigid distinctions, and personified what we now usually think of as natural forces.

Modern adults still continue to personify mammals, birds, and reptiles, and many still include insects in the category of mind possessing phenomena. But most people have given up on oceans and mountains and trees and relegated these phenomena to the category of matter only.

Those who now theorise about the nature of mind in non-theological terms, mostly seem to have concluded that it emerges when biological nervous systems reach a certain threshold of complexity and sophistication. Such Emergentism describes mind as a mere epiphenomenon of matter, rather as we might describe rainbows as a surprising side effect of planetary meteorology. Darwin's theory of The Evolution of Species has lent considerable support to the idea of Emergentism, as it shows a gradual increase in complexity resulting in some creatures which think they have minds.

However a radically different view remains possible. Perhaps mind constitutes a fundamental property of matter, and all matter does mind activity of some kind, and we should not regard it as dead and inert.

Back in the days when thinkers felt fearful of espousing outright atheism, the idea of matter as a living substance found expression in the idea of Pantheism. To a pantheist the universe itself constitutes the mind of god. Every last star and atom constitutes a component of the mind of a god who does not exist separately from the universe which as a whole functions like a living creature, and we can regard ourselves as thoughts within a mind universe.

Gradually the theism leached out of pantheism as it became apparent that the universe did not act as though its mind corresponded to that of some vengeful elderly gentleman with a rigidly authoritarian moral agenda.

The spirit-matter duality merely comprises a moral distinction. If the entire universe consisted of spirit or if the entire universe consisted of matter, then we would have no way of distinguishing which it consisted of, because they would both have to act in an identical manner to produce the universe we perceive. Religions mostly depend on the assumption that the universe consists of good spirit and bad matter and then they further confuse the issue with some bad spirits and some acceptable forms of matter, or at least some acceptable forms of behaviour on the material plane.

So if the thinking pantheist must abandon the theism and seek a strict monist paradigm in which spirit, mind and matter consist of the same phenomena, what does that lead to? It leads to Panpsychism.

Part 2. Panpsychism

Panpsychism has a history. Some anthropologists identify Panpsychic ideas in Animist and Shamanic systems. We can identify Panpsychic ideas of various kinds in the works of many philosophers including Thales in ancient Greece, Cardano and Giordano Bruno in the renaissance, then later in the works of Spinoza and Leibniz and Schopenhauer, and in more recent times in the works of Whitehead[3] and Chalmers.[4]

Panpsychism solved the mind-matter problem at a stroke. If matter naturally includes mind, then the presence of mind in the universe should occasion no surprise nor create any metaphysical paradox, for it occurs everywhere. Panpsychists dismissed the lack of apparent mental activity by teacups, tables and chairs on the basis that either it occurred so slowly that we could not perceive it, or that such phenomena consisted merely of more or less incoherent aggregates of their constituent parts, and therefore do not exhibit much more mental activity than those constituent parts.

However the ubiquity of mind proposed by these philosophers did not find favour with Christian theologians who wanted to maintain a strict spirit-matter separation, and interest in the idea declined from an apogee in the nineteenth century in favour of a mechanistic Emergentism fuelled by the success of Darwinian evolutionary theory.

But then along came Quantum Physics, and after a while it became apparent that the behaviour of the fundamental building blocks of matter and energy did seem to exhibit mindlike behaviour from a certain perspective.

Quantum physics has a reputation for producing contra-intuitive experimental results which permit a wide spectrum of interpretations about what sort of reality underlies them. One interpretation states that no underlying reality exists. This seems less shocking when you consider that quantisation means we cannot continuously divide nature, at some stage we seem to come to the smallest possible bits of reality, and if so, nothing simpler or more fundamental can underlie them, the chain of cause and effect ends there.

In practise the whole universe seems to run a very economical number of types of quanta. Atoms have only electrons orbiting just two types of quark which make up the protons and neutrons in their nuclei. We also have photons which account for light and most other rays and radiations. Two heavier versions of the electrons and the two types of quark do sometimes appear, but they play very little part in the activities of the universe. A couple of other energy exchanging particles seem to make nuclear processes work and the universe swarms with very tiny neutrinos which don't seem to do much except help old exhausted stars explode. The behaviour of this small number of types of quanta leads to all the splendidly complex and peculiar events we observe in the universe.

Quantum Panpsychism depends on the idea that the basic quanta of matter and energy exhibit mind-like behaviour. Both mind and quanta exhibit a mixture of apparently causal and random behaviour.

If we take 'Free Will' as a defining quality, or perhaps THE defining quality of mind, then we cannot explain it satisfactorily either in terms of deterministic or random behaviour, and we seem to have a paradox. Few people like to think that their behaviour always arises as a completely automatic response to circumstance. Few people like to think that their behaviour always generates itself randomly either.

However, on closer inspection of the thinking process, it appears that we actually conjure free will quite satisfactorily from a mixture of deterministic and random mind processes.

If I cannot decide between alternatives because each has equal logical or emotional appeal, then I end up choosing randomly or by mere whim. If no alternatives suggest themselves in a situation then I allow ideas to arise and combine randomly until I find something that makes logical or emotional sense.

In practice I actually use a complex and stratified mixture of these procedures to reach decisions. Free will would have no use if it meant absolute freedom from all previous conditions and the demands of current circumstances.

Thus by using a mixture of deterministic and random processes I arrive at decisions which lie within limits but which no agency, including me, could predict with certainty beforehand. I submit that what we call free will consists precisely of this kind of activity.

If someone claims to have free will, ask them, 'free from precisely what?'

We could fairly easily build information processing machines which exhibited any degree of free will by using the above principles. However we usually prefer to aggregate machines to do exactly what we want. When they act unexpectedly we tend to get annoyed with them.

Chapter 5 presents evidence for the irreducible 'randomness within limits' in the behaviour of the quanta underlying reality, but for now it remains assumed.

Although quanta have a simple form of free will, because they behave randomly within limits, most forms of bulk matter behave fairly deterministically and we can describe their behaviour with the approximation of 'cause and effect'. This arises because of the law of large numbers. Throw one dice and any of the six numbers may come up top, but throw six million of them and you will get almost exactly a million of each of the six numbers. The total of all the top numbers showing thus always comes out to almost exactly three and a half million every time. The more dice that you use, the smaller the deviation from exactly a one in six appearance of any number becomes.

Random quantum behaviour can thus lead to apparently causal macroscopic behaviour.

Large aggregates of quanta such as billiard balls thus behave predictably and with apparent determinism for short time periods.

Yet if bulk matter aggregates or acts in such a way that some of its component quanta can affect the behaviour of the whole, then that whole begins to act with free will. The weather acts like this, and so does the brain. Even a 'low-minded' billiard ball exhibits non-causal behaviour eventually. The final position of a billiard ball becomes progressively less determinable in advance as it undergoes more and more sequential collisions. If it sets off with enough momentum to bounce off the cushions of the billiard table more than about 7 times, then its final position remains indeterminate until it happens. We can calculate the limits of this indeterminacy, and they equate roughly to the entire area of the table, so the ball could end up anywhere on it.

Some philosophers regard Panpsychism, the paradigm of the ubiquity of mind, as neither provable nor falsifiable, and therefore that it lacks use or consequence, and thus that it merely qualifies as a mystical belief system.

However quanta do exhibit a number of behaviours that do not always appear on the macroscopic scale of tables and chairs and stones, and these seem far more mind-like than the matter-like behaviours we get used to on the macro-scale. In particular, under certain circumstances, quanta seem to 'remember' what happened to them, and they also seem to 'communicate' with each other without apparent material contact.

(Chapter 5 deals with these phenomena of 'quantum weirdness' in some detail.)

Such quantum activities may explain how the apparently 'material' brain performs apparently 'mental' activity and why parapsychological events sometimes occur.

Quantum Panpsychism can perhaps give us an economical explanation of how magic occurs and also provide some ideas on how to improve its effectiveness in practise.

Part 3. Quantum Panpsychism and Magic

In a dualistic spirit-matter or mind-matter paradigm, any kind of mind to matter effect (including ordinary thinking) appears mysterious, or parapsychological. Matter to mind or matter to spirit effects remain equally incomprehensible, or even more so if you put spirit in some sort of superior position.

Now spirit-matter dualists frequently cite miracles as evidence for the reality of spirit or spiritual agencies. Claims of miracles underpin most religions, and most religions have a habit of interpreting the most trivial anomalies as hard evidence.

Non-religious magicians tend to regard parapsychological events as evidence of nothing other than magic, because they can occur in non-religious contexts and also in the contexts of religions which specifically deny each others validity.

Any religion which considers another religion false finds itself in the ridiculous position of having to attribute any miracles manifesting in the other religion as arising from the activities of the devils in its own.

Quantum panpsychism suggests that we turn the whole argument on its head and interpret parapsychological events as evidence for the absence of spirit or mind as phenomena separate to matter.

Miraculous, parapsychological, magical events tend to occur rather capriciously and infrequently on the macroscopic scale. However on the quantum scale they occur frequently and in a much more dependable fashion. The quantum level of reality seethes with weirdness, quanta appear to teleport by disappearing at one place and appearing at another, they appear to communicate instantaneously across space and probably time as well, sometimes they appear to exist in two places simultaneously, or in two contradictory states at the same time, and they may travel backwards in time.

Thus we have a case for recognising the quantum level of reality as the real home of magical phenomena and the source of what we call free will. When bulk aggregates of quanta become configured in a suitable way, then the phenomena that we conventionally call free will, mind, and magic, can appear on the macroscopic level as well. When quanta aggregate in such a way that their individual weird and random behaviours tend to cancel out, then we observe the causal behaviour that we associate with 'inert' matter.

On a practical level we know that magic, as a deliberate human activity, works far better if we deploy it against phenomena that retain some of the behavioural fluidity of their component quanta. Influencing the weather, or another human's behaviour, or the fall of well thrown dice, gives better results than trying to split stones with your bare unaided brains, although moderate sized pieces of glass sometimes yield to this.

(Glass often contains cooling induced stresses, which leaves it susceptible to both spontaneous fracture and to poltergeist type activity from those with a talent for acute anger gnosis.)

In this chapter I have attributed mind-like behaviour but not 'consciousness' to quanta, and a degree of mind-like behaviour to all phenomena composed of quanta, (and hence to all phenomena). I have no grounds for attributing 'consciousness' to the quanta, but I have no grounds for attributing it to myself either.

Chapter 3 addresses the reasons for this.

Chapter 3
Multimind - Psychology

This chapter deconstructs the superstitions of Consciousness and Self, and seeks an Apophenia in the paradigm of the Multimind Randomaton.

Part 1. The Myth of 'Consciousness'

Consciousness always has a subject other than itself. It always has a focus on some perceptual phenomenon or on some internal state or emotion or thought.

Descartes proclaimed 'I think therefore I am'. Other people may rely on consciousness of different phenomena to reassure themselves that they still exist, but toothache provides almost everyone with unarguable confirmation of their existence.

We cannot however have content free consciousness. It does not exist as a state of 'being', it consists of an activity, and this activity ceases under anaesthesia or deep sleep.

Try as hard as you like with meditation or sensory deprivation but you can never achieve pure consciousness, although you may achieve an interesting consciousness of your own blood circulation or endocrine functions, or of some mystical feelings or ideas.

So how does the subjective impression of consciousness as a state of 'being' arise?

Look again at Descartes' assertion of 'I think therefore I am'. The appearance of the word 'I' twice gives the game away. Plainly the two instances of 'I' cannot refer to the same

Theriomorphic atavisms of the Multimind.

phenomenon. Descartes must contain an 'I' doing the thinking, and an 'I' observing the other one doing it. Any form of introspection implies a dialogue of some kind.

Plainly we should regard 'mind' as a verb, as an activity of the brain, rather than as a 'thing' which we have, or consist of. Mind remains unobservable; it consists of a doing, not a state of being. We can only infer the presence of the activity of minding.

Consciousness only occurs when it has a subject, so self-awareness can only consist of one part of the system having awareness of the activities of another part. However we learn to assume that The Same Part always has consciousness of the rest.

We probably have to adopt this assumption to retain a sense of personal coherence as a survival strategy, even though the evidence all points in the opposite direction.

Writing in a book of short essays about things we believe but cannot prove,[5] one neurophysiologist quipped that he believed consciousness works as a sort of trick we involuntarily play on ourselves, but that understanding the trick might send us all to hell. Buddhists philosophers might argue that such an understanding could set us free.

The Philosophical Zombie describes a creature in a famous thought experiment.[6]

This hypothetical Zombie has all the usual attributes of a human except that it does not have our subjective conscious experience of events but acts entirely on reflex like a massively sophisticated automaton. Thus it withdraws from stimuli that its programs consider harmful, and it seeks food and water and reproductive opportunities and so on, as its programs compel it to. It can also make what sounds like perfectly intelligent conversation and pass the Turing test with flying colours, but it has no 'consciousness' even though it can monitor its environment and its internal states.

We would almost certainly have to make such a massively sophisticated automaton using organic chemistry, so it would consist of meat rather than metal, just like us.

Some theorists tend to conclude that such Zombies could exist and function without consciousness, so perhaps consciousness doesn't really exist at all except as an illusion. Perhaps we simply have to delude ourselves with a fictional sense of consciousness to create a sense of simple coherence inside an otherwise impossibly complex information processing device.

Others think that such a Zombie could not exist or function convincingly as human; because real humans require something qualitatively different called consciousness. They conclude that such a creature would behave more like a science fiction android automaton. 'My senses inform me that my foot has started burning, I shall therefore remove it from the source of heat in accordance with my survival imperatives'.

The creature would appear to lack what we call the subjective conscious experience or 'qualia' of pain. It seems unlikely that any degree of response sophistication could completely disguise this, even if we built in an automatic scream.

I beg to differ with both camps. I suspect that a creature with only a single consciousness would behave like the automaton type of zombie, and that we cannot understand consciousness if we assume that we have it in singular form only.

In the course of normal everyday life the assumption of singular consciousness works well enough, but in extremis we see a different picture. Consider the 'qualia' of pain, it behaves as though it consists of an independent 'pain consciousness' and as it becomes more active, our other consciousnesses start doing less and less, the pain consciousness becomes dominant, and you find yourself observing yourself mainly from the perspective of pain.

People who practise extreme forms of meditation or concentration or mystical activity report that their consciousness of everything else decreases. Normally people tend to identify the consciousness that they perform as 'their own', but they may afterwards disavow extreme states, and claim that they came from elsewhere, particularly from spirits if they have religious inclinations. Many creative people claim that their inspirations come from a source that they do not identify with their normal consciousness. Their normal consciousness has awareness of the other source but does not seem to include it. But conversely, when the other source becomes very active, normal consciousness can become a subject of its observation, but eventually the other source may cease to notice the increasingly inactive normal consciousness.

Anger provides a simple example of this. When one feels anger rising, the normal consciousness has awareness of the increasing activity of the anger consciousness, and vice-versa. For a while it may remain in the balance which will become the most active and which will mainly observe the other. In extremes the anger consciousness may enter into a dialogue with body consciousness instead, whilst the normal consciousness shuts down. Afterwards, people who rarely experience such states may find difficulty explaining or remembering their actions in normal consciousness, they may even disclaim agency in terms of diminished responsibility.

Consciousness has the odd subjective property that it seems to have the ability to flit from doing one qualia or state to another, and often of doing several at the same time. All this does seem paradoxical if you insist on having only a single consciousness, the 'me' or the 'I'. On the other hand if we assume that all 'our' qualia and states exist as separate consciousnesses, then it makes considerably more sense.

From a quantum panpsychic perspective it appears impossible in principle to construct a philosophical zombie because any sufficiently complex information-processing device that can monitor its environment and its internal states will inevitably have consciousnesses well before it has a processing power equivalent to the human brain. At the time of writing, computers hardly exceed insects in their processing power. If we wanted to build a device that convincingly mimicked human responses we would have to endow it with many separate programs that competed for control; and which to some extent monitored each other. Each of these programs would inevitably have consciousness to some degree.

The quantum panpsychic view endows all phenomena with a degree of mind-like behaviour anyway, and quite modest quantities of brain tissue can support extensive monitoring and control programs. The human brain weighs about as much as the brains of 45 cats, or 700 rats, or an astronomical number of insect brains. We know that many parts of it have highly specialised functions. The human brain actually supports many consciousnesses. Some of these become active only infrequently, some monitor the activities of some of the others, but probably none monitors all of the others. A conspiracy of the more active consciousnesses usually learns to define itself as 'consciousness in the singular' in monotheist and post-monotheist cultures. We learn to regard ourselves as 'individuals' despite that we have profound internal divisions, and we have to make big efforts and sacrifices to create a unitary sense of self. In magic and mysticism and in creative thinking, we can gain much by relaxing the grip of the unitary consciousness that we have learned to construct. Part two of this chapter deals with the construction of self, and part three deals with undoing it.

Part 2. Constructing the Self

The Self arises largely as a social construct. We become assembled from bits and pieces of other people. We start by receiving genetic material from our ancestors and then we go on to receive language and ideas and behavioural patterns from our parents, peers, and teachers. As we age we seem to develop some ability to choose what to incorporate into ourselves, and we select various add-ons available in the media of our culture.

At an early stage we seem to somehow develop 'theory of mind' as we come to the realisation that other people have 'intentionality' and act somewhat differently to say, refrigerators. We arrive at the idea that other people have minds which may lead them to behave as if they had intentions and concealed agendas. Autistic people may owe their condition to an impairment of the ability to develop theory of mind.

In the normal course of development, theory of mind attributes a single mind to each significant other person. However if a significant other behaves in wildly differing and contradictory ways it can lead to eccentric and possibly dysfunctional ideas about self and others in general.

Gradually we begin to apply theory of mind to ourselves and learn to recognise various intentionalities within, and we also learn to deceive and to lie. We come under intense pressure to conform to consistent behaviour patterns. Parents and teachers pressure and intimidate children continually in various subtle and sometimes not so subtle ways to exhibit approved behaviour, and then express surprise if they bully any of their peers who exhibit any sort of differences.

As a social species we exhibit an extraordinary suggestibility. It takes a chimpanzee about six years just to learn how to break nuts with two stones, in the same time a human has learnt half a language, a large suite of complicated physical skills, and the beginnings of a system of beliefs about the world.

We also learn to present a fairly consistent self to the world. Out of character behaviour attracts disapproval or punishment. Nothing instils a belief more strongly than persistently acting out the behaviour that goes with it. We do not so much do what we believe, as believe what we do. Quite soon we internalise the idea of the singular self because our culture demands that we act as though we had one.

For further commentary on this kind of view of the nature of mind see the work of Norretranders[7] and Ornstein.[8]

The singular self remains a defining feature of monotheist and post monotheist cultures. It confers a greater sense of personal responsibility than our pagan forebears would have felt comfortable with.

Every theology, pantheon, and demonology implies a psychology. Most pagan cultures attempted to include a wide spectrum of possible selves and behaviours, with a god or goddess or a minor deity for just about any activity, allowing them to make love or war or whatever, as they felt the inspiration to do so. Thus they seem to have thought and acted with less of a sense of internal conflict and less of a sense of personal agency than we find normal today. Thus violence and unrestrained sexuality seem to have featured as everyday phenomena in many early pagan cultures, rather than as occasional paroxysmal outbursts as they do in ours. As pawns of the gods of their own creation, the pagans gave themselves licence to express their impulses and selves to the full, especially if they occupied a position in society that gave them the power to do so.

However city life threw up many challenges to later paganism. Increasingly complex rule structures evolved to cope with the expression of pagan impulses within densely packed populations, and pantheons tended to proliferate rather absurdly as the Romans in particular attempted to incorporate cults from all over their empire. It seems likely that the majority of notable Greek and Roman thinkers paid only lip service to their official religions, but we owe the ideas of the muse, the daemon and the genius as quasi-independent sources of personal inspiration, to these cultures.

Monotheism certainly brought a brutal simplicity to the questions of social control and personal behaviour. Half of all behaviour got defined as approved by the single deity, and the other half got defined as damned. Monotheism mounted a two pronged attack on pagan cultures. It appealed to the rulers of societies as a superior means of social control, (they usually considered themselves above the moral precepts anyway), and it appealed to the poor masses as it made a virtue of avoiding the sybaritic excesses that they could not usually afford to indulge.

Monotheism brings with it an increased sense of personal agency and individual selfhood defined by the supposed free will to choose between what god and society requires and what personal impulses suggest. In monotheism you cannot always find a god that agrees with you, so the daemons that inspired the pagans become the demons that culture now expects you to reject as not-self. This creates a thriving industry of self-loathing and guilt. Monotheists define themselves at least as much by what they don't do (or pretend not to do) as by what they do. Expect extensive lists of prohibitions from any monotheism or post monotheist secularism.

The post monotheist westernised democracies have largely retained the paradigm of the mono-self and refined it in many ways. Secular law now attempts to both reflect and lead belief as religious based law once did. You can believe more or less what you like so long as you don't express beliefs critical of certain other classes of people, but intense social pressure falls on those whose beliefs or actions do not conform to certain standards of self-consistency.

Whilst a wide range of roles and hobbies remain available, our culture regards many as exclusive of certain others. Consider this short selection:

Astrologer, Politician, Priest, Scientist, Prostitute, Schoolteacher, Businessperson, Druggie, Artist, Police Officer, Model, Lawyer, Magician, Soldier, Erotic Novelist.

Whilst many people could easily have any of these activities as a career and another as a sideline or hobby, the social conventions of consistency usually discourage or prevent many possible combinations, for few discernible logical reasons whatsoever.

But don't we find it fascinating to discover someone who has two 'incompatible' identities?

The word schizophrenia comes from the Greek roots 'divided' and 'mind' and in the popular imagination it often means someone with two minds, at least one of which seems mad. An old joke puts it thus, 'when a man speaks to a god its prayer, when a god speaks to a man its schizophrenia'. In psychiatric terms schizophrenia covers a very poorly defined group of maladies that does not invariably include hearing voices, although this symptom frequently provokes that diagnosis.

Many people hear voices without suffering any of the debilitating and dysfunctional effects associated with schizophrenia, some treat these voices as sources of inspiration or develop religious ideas about them, others become mediums or occultists.

The idea of demonic possession occurs in most monotheist cultures but post monotheist paradigms usually describe it as some variety of schizophrenia. Yet possession sometimes gets treated as a desirable state to achieve, as in the Voodoo faith or in some other ecstatic cults.

Despite its popularity in pop-psychology, Multiple Personality Disorder very rarely manifests in its recognised psychiatric form where some of the selves have complete amnesia about the activities of others. It would seem that anyone can present a different persona in different circumstances, but that severe trauma can induce a permanent split between those personae.

The classical psychological concepts of the unconscious and the subconscious minds arose in a culture that expected people to act in a considerably more reserved and repressed fashion than seems normal today. Sharp divisions between the conscious, the subconscious, the unconscious, and perhaps the super-conscious (whatever that may mean), now appear rather artificial and contrived. Some memories, thoughts, emotions and impulses merely acquire more of a propensity to take control of the whole organism than others. Many of them operate without much direct communication with what the early theorists called the 'ego'; another rather loose concept derived from the Latin word for 'I'.

The fact that the mind tends to produce confirmation of any descriptive scheme that we impose on it, including the Freudian Id, Ego, and Superego or the Kabbalistic Sephiroth of the Tree of Life or the Eight Circuit Wilson-Leary model, surely tells us something. No part of it can comprehend the whole incredibly complex and malleable assemblage.

All in all, it seems that humans can function across a whole spectrum from the apparent Mono-Self type to the Multi-Self type. In practise neither extreme of the spectrum seems optimal, because at both ends of it the selves erect barriers between each other.

The Mono-Self type acts predictably and with restricted creativity, and has a cellar full of demons and discarded angels. The full-blown Multi-Self type can act creatively and unpredictably, but erratically and dysfunctionally if communication between the selves breaks down.

We need to aim somewhere between the Zombie like automaton of the mono-self type and the disintegrated condition of the complete Randomaton to explore the multitudinous riches within and to emerge in a functional and sane condition.

Monotheist mysticism and magic inevitably plunges its practitioners into the demon realms.

Monotheist mystics exalt one imagined god-self within by repressing all their own natural ungodliness. They never succeed in this until perhaps old age erodes their sexuality and aggression and appetites, but in the meantime they sometimes manage to sublimate their impulses into 'good' works. But expect outbreaks of appalling behaviour or long nights of unproductive guilt and anguish at the very least.

The Devil gave his name as Legion, the legion of repressed selves lurking in the monotheist's dungeons.

Part 3. Dicing with the Randomaton

Chaoists approach multi-self management with stochastic techniques. If one self doesn't work, try another; if necessary, at random. Here we see lateral thinking at work on the grand scale.

Most people seem strangely protective about their name and immediately correct you if you so much as mispronounce it. On the other hand, in many mystical organisations people often have a special name which they only use within it. A change of name or title seems charged with considerable significance for most people. I once spent a year and a half in a job where they called me Jim rather than Pete, due to someone mis-hearing something on the first day. I decided not to disabuse them. It worked out rather well, Jim did a better job of educating the unwilling and the behaviourally challenged than Pete would have, and Pete refused to take Jim's identity and job home after hours.

This seems to work best where you can enter a new situation. Asking everyone you already know to call you something different has little effect in the short term and gains you no extra degree of freedom.

Apparently everything perceived in our universe has a name, and whenever anyone comes across something lacking a name they seem to feel an overwhelming compulsion to give it one. Yet in bizarre contrast to this, few people have any names at all for any of their many selves. Half of their universes consist of murky areas full of phenomena that don't even have proper names. Mere psychological tags often have to suffice, even for the relatively self aware.

Despite that we can peer into the hearts of stars and atoms; our psychology remains primitive. Arguably we have little more real psychological knowledge than the ancient Greeks did. The destruction of all books on psychology would have no serious consequences at all.

Naming the selves of the personal mythos might seem like the first step on the road to insanity and the disintegration of the ego or self image, and we might well ask 'who' names them. In the absence of any sort of 'real' inner core or 'essential self', the selves have to name each other or at least to exchange names and welcome each other to a party that has no host with special privileges, because they all own the building.

I tend to favour democracy, it looks like the least worst system of governance yet devised. Critically, it depends on all power blocs allowing other blocs to try anything that does not radically obviate their own agenda. It does not work in highly divided 'societies'; it depends to a large extent on negotiation between various interest groups.

A truly sane individual or society tries to achieve a compromise between all its impulses.

We (the author) have endeavoured to conduct our life as a party, with something to amuse and exercise the skills and obsessions of all those present at various points during the celebration. In the absence of an adequate psychological terminology we have tended to identify each other with the names of the now safely dead classical gods from various pantheons.

Take violence for example. Everybody has a self that loves violence, whether they try to repress it or not. Don't pretend that several million years of evolution has not equipped us with a certain facility to relish hunting, fighting, and killing, and the crushing of rivals and enemies, and given us a sense of glory and achievement in doing so. However a Mars self unadvised by our other selves, leads the whole organism rapidly to disaster.

Plus of course people don't generally like anyone manifesting a Mars self except under the controlled circumstances of sport or entertainment. Watching violent sport and entertainment seems rather like watching pornography and then not having any form of sexual activity. It titillates an impulse but does not satisfy it, and it allows the maintenance of the hypocrisy that we abhor violence. In fact we have a self that loves violence and several others that don't like it, and they usually have a bad opinion of the self that does. Thus the violence presented in entertainment for the viewer to identify with usually has to appear as justifiable revenge, anything else seems immoral to several of the other selves.

We* (the author*) let Wotan, as we call him, out of his cage for regular ritual exercise. He likes weightlifting, sword practice, the thunderous roar of drums and cannon, the crash of axe upon shield, fire, explosions, muscle powered projectiles such as javelins, knives, arrows, etc and getting into an ecstatic rage for the hell of it. Well why not?

Anger seems a much-neglected resource. It can temporarily double your physical strength and concentration during really hard work, it can project a sort of madman-charisma that wins conflicts psychologically, and it can also serve as a gnosis for projecting intent magically.

We* (the author*) don't feel ashamed of Wotan, we can trust him not to act out of turn, we regard him as a valuable committee member, he likes devising and playing complex board wargames with Logicus the abstract thinker, which neither of them would probably enjoy on their own. Wotan regards ordinary individual human stupidity as rather laughable and only gets aggressive at organised stupidity and malice.

Then we* find that we also comprise at least half a dozen other Selfs with various agendas and abilities, and that all of them seem to have magical powers if the others will stand aside for a while and let them do their stuff.

Death provides constant saturnine advice on matters of time, ageing, senescence, mortality and futility. Sex seems more polymorphous-perverse than the rest of us realised, and has developed a delightful repertoire of fairly harmless paraphilias over the years. Love appears as several different characters that love quite different phenomena, and get quite different payoffs for doing so. The same goes for Hatred. This realisation solved an awful lot of confusion and argument. Logicus would no more try to rationalise any of our Hatreds away than he would try to kill any of our Loves.

So which of my who's am I?

We* regard that question as meaningless because it contains a false imputation of 'being' in the use of the word 'am'. We* have no chairman at our round table, the microphone gets passed around according to circumstance or purely randomly if no circumstances impinge. If we* have any kind of real or fundamental self it consists of the quantum panpsychic chaos underlying all of our* consciousnesses. The Ancient Greeks considered that their gods arose from Chaos, they had a point there.

Great people invariably contain great contradictions, internal self-consistency has no virtue, it merely causes mediocrity. Rather we should strive to make the most of all the selves that we contain, for each can function as a god for a time if the others stop trying to restrain it. We* seem to function better by regarding ourselves as a team, and by occasionally letting one of our number manifest in full god form, but more of that in Chapter 4.

*Some Chaomeras of the
Neural Neopantheon;
We have worlds within us
And we have others within us
Humans and gods make each other
In each others images*

Chapter 4
Neopantheism
- DIY Religion

This chapter looks at possible ingredients for non-insane DIY religion. It begins with a demolition of the whole idea of objective truth in theology and seeks an Apophenia in the Neo-Pantheist concept of a personal mythology and narrative.

Part 1. Against Logos, 'The Literal Word'

Some people have a mystical capability. They can find awe and wonder in the natural world or in the astonishing phenomena of consciousness itself, or simply in the fact that they, or indeed anything at all, or anyone else, actually exists. Others only seem to have a religious capability. They just want some answers to the big questions to believe in, and they will accept any absurdity rather than uncertainty.

Of all our instincts the religious one seems particularly vulnerable to our profound suggestibility. All too easily it gets subverted for the purposes of social and political control, or simply to make a living for wicked old men.

Most of the religion that litters our planet seems indistinguishable from mental illness.

It blinds people to the enormity and variety of the universe and themselves, it tends to narrow rather than to expand horizons, it takes myth and metaphor for literal truth, it values faith over evidence, and it seeks to impose certainty where open mindedness has more to offer.

If any individual in isolation developed a series of beliefs and behaviours equivalent in their irrationality to most of the main religions, everyone else would regard them as deranged. Let's try it:

How about a prophet or a messiah born from the anus of a man for a change? That sounds like a suitably impressive and contra-intuitive miracle. The great Sky God sent his emissary to us by this means to remind us that He creates universes out of black holes. Devotees must of course carry a symbol of the sacred 'O' ring at all times. A whole elaborate morality thus depends on the correct and incorrect uses of the anus. On feast days we celebrate its functioning, on fast days its functioning becomes punishable with burning stakes. On judgement day only the worthy will squirm through the great black sphincter in the sky, but the rest will spend eternity in a great boiling sea of, - well I guess you can fill in the theological details.

Of course this sounds deranged, yet it has about as much coherence as any organised religion, and when millions of people come to believe in it we will have to respect their beliefs or they will become very angry and probably very violent if they gain secular power. Anusites will crush the unbelievers, apostates and blasphemers!

Indeed they will take a dim view of anyone who rejects The Word of the Black Hole.

We can never know for sure in what sense the ancients believed in their gods. Did they believe in Logos type gods that really existed in some objective way as actual independent entities, or did they believe in them in the Mythos style, as metaphorical principles to explain the world and the human heart?

The belief mode of the ancient Egyptians remains obscure because their hieroglyphs do not submit to unambiguous interpretation, and they seem to have lacked the vocabulary for abstract thought, as we know it. Perhaps this in itself provides a clue as to how they thought. Mythos and Logos seem indistinguishable in what we can make of their inscriptions. Maybe they lived and breathed and thought entirely in one mode and expressed themselves exclusively in mythological terms. We often forget that the religion(s) of ancient Egypt spanned millenniums and a huge serpentine territory. Individual ancient Egyptians would only have venerated a small selection of the gods now known to us.

The classical Greeks however present a different picture. Plato made a clear distinction between logos and mythos style thinking and it seems likely that the majority of noted thinkers in ancient Greece probably regarded the myths and stories of the gods as metaphorical truths and explanations rather than as actual literal truths.

The peasantry however may have taken such tales literally but in small doses particular to certain areas only. The entire classical Greek pantheon looks like a huge family tree of fornicating and squabbling deities with ever more ludicrous stories attached, and surely no scholar familiar with too broad a swathe of it could have taken it all at literal face value. The flowering of abstract non-mythological thought in the golden age of Greece, which contributed so much to art, mathematics, philosophy, politics and science, could hardly have come about in a culture dominated exclusively by mythos style thinking. When the ancient Egyptians discovered something useful by accident the knowledge invariably became incorporated into their mythology. If the ancient Greeks discovered something by experiment they often allowed it to stand on its own as a non-theological idea.

Roman civilisation represents a bit of a setback in many ways. It took the Greek religion on rather uncritically and it failed to adopt many of the insights in Greek philosophy. Disastrously it failed to adopt Greek mathematics although it still managed to build an awesome bureaucracy and hence an effective army filled by state equipped peasant levies rather than by self equipped aristocrats.

Historians advance many reasons for the collapse of the Roman Empire. Undoubtedly it suffered from imperial overstretch, dynastic power struggles, and military problems with barbarian cavalry, but it also ran into severe religious and philosophical problems. The Romans attempted to amalgamate the religions of conquered peoples with their own, and as Rome became more cosmopolitan it imported foreign cults wholesale. The cult of Mithras became popular in the army; and cults of Isis appeared in the cities. Rome itself ended up swarming with the priesthoods of various deities along with every kind of soothsayer, diviner, prophet and magician.

Out of this confusing and increasingly incredible stew of paradigms one particular religion of Hebraic origin evolved to eventual dominance and then eliminated all opposition with an iron fist. At the Council of Nicea 325AD the empire set its beliefs in concrete forever. Before that, huge differences of opinion existed between various vaguely Christian groups around the empire.

Only one god existed. It created the entire universe. It required worship. It required obedience. All other religions were wrong. Mythos style thinking ends here with the adoption of the Hebraic idea of the literal and absolute objective truth of a written religious corpus.

At the Council of Nicea the assembled worthies decided on exactly which written texts would constitute The Truth. They had plenty to choose from, and they had to discard most of the material available to them.

This stood in violent contrast to paganism which had no absolute texts at all, but had oral or written stories which it could elaborate on or alter or interpret according to taste and usefulness.

One might argue that the Roman Empire never really fell, it merely switched from mainly military to mainly religious methods of control and within a few hundred years it actually controlled more territory by the latter method.

The new Logocentric monotheism with its insistence on the literal truth of The Word of its scriptures not only discouraged mythological thinking, but it also discouraged reasoned enquiry into any other form of truth but its own. Logos in the sense which Plato intended it, the enquiry into reality by reason, lay dormant for centuries, a period which we now call the Dark Ages. During that period another intensely Logocentric monotheism arose in the Arabian Peninsula and it used exactly the same technique, a Sacred and Absolutely True book.

It took Christendom many centuries to begin to extricate itself from the idea of a fundamentally true logocentric religion and start to apply reasoning to the natural world instead of theological matters. The process seems to have begun in the renaissance with the rediscovery of Greek ideas. The invention of the printing press sparked off the reformation which helped a bit, but the Enlightenment took a long time coming. Even today some people in westernised nations seek a retreat into fundamentalism whilst many cultures of the third major monotheism remain mired in it.

Note that Logos style thinking underlies both the idea of literal truth in religion and objective truth in the material world. The results of Logos style thinking depend on whether you apply it to belief or to observation, and so do the results of Mythos style thinking. We can arrange these ideas graphically to see what paradigms result:-

Figure 1.[9]

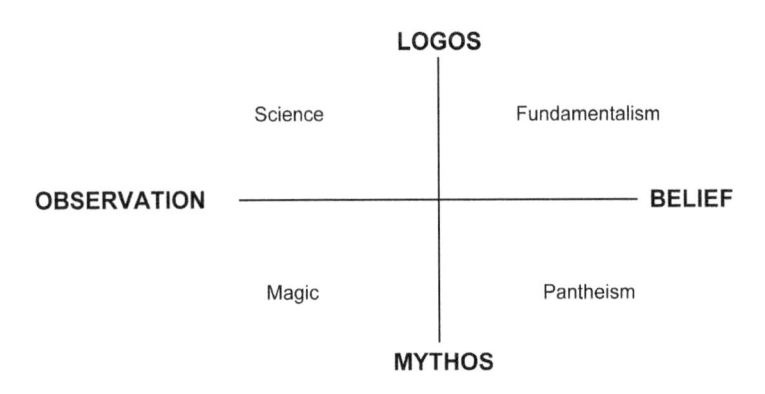

```
                        LOGOS
                          |
           Science        |    Fundamentalism
                          |
OBSERVATION ──────────────┼────────────── BELIEF
                          |
           Magic          |    Pantheism
                          |
                        MYTHOS
```

The terms 'Magic' and 'Pantheism' have a rather looser and more inclusive usage than normal in this scheme. Magic includes more or less any attempt to use mythos style thinking about the observed phenomena of the world and it thus includes astrology and alchemy. Pantheism refers to the mythological/ analogical attitude to belief and could in theory include polytheism or monotheism. Note that Fundamentalism can include polytheistic fundamentalism as well as the more common monotheistic fundamentalism.

Figure 1 represents a graph, and various schools of thought can occupy areas anywhere in the quadrants

My average compatriot in these British Isles has a paradigm footprint or 'psychogram' consisting of a blob centred roughly on the origin where the axes cross.

Such a hypothetical person has a general feeling that an objective reality open to rational analysis actually exists (Science). Nevertheless this person has a vague intuition that fate and intent can play a part in life (Magic). Notwithstanding this, such a person has a head full of archetypes, celebrities and narratives (Pantheism). Lastly, when it comes to the big questions of life, existence, and death, the average person usually maintains that 'There Must Be Something' (Fundamentalism).

Other cultures and individuals and schools of thought will obviously have quite different paradigm footprints or psychograms on the figure shown.

Chaoist philosophy in general, usually has an epicentre focussed on the lower left quadrant. It regards existence as basically random and chaotic but subject to the possibilities of Psychic and Physical anticipation and manipulation, and to manipulation by Belief. Thus it has tendrils extending into the Science and Pantheism quadrants. Chaoist philosophers conspicuously avoid the upper right quadrant, the domain of the Sky Fairies, the mainly monotheist gods and devils, and the whole associated plethora of other 'literally real' spirits.

The Sky Fairy quadrant differs from the others in that faith alone maintains its paradigm in the absence of evidence. Science either makes material things happen, or gets it wrong. Magic either gives useful results or it doesn't. Pantheism either supplies an agreeable narrative to live by or it fails to do so.

Fundamentalism on the other hand makes a virtue of contra-intuitive and contra-evidential faith. Indeed, only irrational beliefs can actually work for a 'literal' religion because people will not make emotional investments in defence of perfectly obvious truisms, only in defence of highly questionable ones. Faith exists only in the context of a continual internal dialogue with doubt.

Favourite topics for contra-evidential faith usually revolve around such absurdities as that you will live happily for ever whilst bad people will get their just deserts in eternal hell, and that you will get all the things you wanted in this life but didn't get, after you're dead.

Faith needs to fail to deliver the goods most of the time to attract investment of thought and emotion in it. Faith abhors blasphemy and fears apostasy because these raise those very doubts which the faithful spend so much time suppressing with ritual and prayer. Prayer basically consists of talking yourself into believing something you understand as rationally false, and then asking it for the occasional favour.

So where does the widespread idea of literally real gods and spirits come from?

It comes from the same 'theory of mind' facility that has evolved to equip us with a working hypothesis about the existence of minds in other people, (and animals), and a self-image.

Do other people actually exist? Well they exist to the extent that we either invite them into our heads or they manage to force their way in. Friends, family and colleagues may have more reality for us than people that we have not met, but politicians, celebrity figures from the media, characters in novels and comic books, people appearing in dramas and entertainment, personal heroes, all these have some sort of existence for us. Note the deliberate mixture of fake and genuine, real and imaginary, and dead and alive characters here. I describe anyone I've not actually met as 'imaginary'. (Only lunch can translate imaginary people into real people.)

Out of such experiences we build our own identities by a process of dialogue and accretion. We listen to real people and absorb their attitudes and mannerisms but we also do this with 'imaginary' people in all the various media of oral stories, art, theatre, books, radio, film and television etc. Afterwards as we reflect on our experiences of real and imaginary people we find ourselves using theory of mind on them and they acquire a reality of sorts inside our own heads.

Unfortunately our suggestibility can easily derail this highly useful ability, particularly when the suggestion gets applied heavily in youth with the full force that a culture can bring to bear. For much of history people have grown up with alarmingly large parasites living inside their minds, Monarchs, Emperors, Gods, High Priests, Dictators, and Gurus.

Unsurprisingly all of these characters have striven to control the media of the cultures in which they live. They want precise control of their own personality cult, and they don't want any competition. The growth of uncensored and uncontrolled media has done a great deal to weaken the hold of the major parasites on people's minds in democratic countries, but elsewhere, tight control of the media has strengthened it.

In a relatively free country you can fill your head with a vast selection of real and imaginary people with radically different identities, and end up with a much larger self image, or you can retreat into dialogue with something simpler like a single god or personality cult figure. In many traditional cultures and in some recent and contemporary hard-line religious or political states, you either believe in the god or demagogue or suffer serious consequences.

Perhaps for the first time in history we live in a world where a substantial fraction of humanity has freedom of belief, and hardly knows what to do with it.

Some adopt a fundamentalism or a single-issue cause or creed to create self-definition, others just seem to wander around lost in the cosmos with no metaphor for self, squandering their belief on one fad or fashion after the other in postmodernist style. Some seem to define themselves entirely by their relationships to other people, and to consist of nothing internally. They have to remain constantly engaged either socially or with 'imaginary' people from the media, or they practically cease to exist in their own minds.

As one exasperated monotheist observed, 'when people cease to believe in god, they will believe in anything', but this begins to look more like the solution than the problem.

Postmodernist, Post-monotheist culture has yet to formally explicate its ideal spirituality, although we can observe many preliminary attempts to achieve this from the New-Age movement, to Neo-Paganism, and Chaos Magic.

Despite their varied degrees of emphasis on transcendence, philosophy, and occultism, all three of these new traditions exhibit a strong current of Neo-Pantheism.

As advanced cultures pass out of a monotheist aeon rendered untenable by scientific thought, and as atheistic or nihilistic scientific positivism and modernism become progressively more questionable, Neo-Pantheism takes their place as the spirituality of choice for the dawning Fifth Aeon.[10]

Both Fundamentalism and Science have started to develop a profound and vitriolic hatred of Neo-Pantheism, and in doing so they have helped to define it. We can take that as a sure sign of the threat that it poses to them both.

Historically, the word Pantheism has covered a variety of beliefs,

That some sort of divine force manifests in all things,

That various gods and spirits pervade all aspects of the universe,

That god remains indistinguishable from nature, and does not consist of a person,

That the universe as a whole has consciousness, or life, or something like that.

Thus Pantheism has a long history, and it has tended to shadow orthodox thought as a species of mysticism for millennia. The emerging Neo-Pantheism of the fifth aeon has many manifestations and little orthodoxy, but nevertheless it has a number of recurrent themes which reflect its Mythos style of Belief. Perhaps it will eventually replace most existing religions. It certainly looks like a spiritual product that has evolved to meet contemporary needs.

Part 2. Neo-Pantheism

At least eight themes seem to characterise the emerging Neo-Pantheism.

I will present them here in their most extreme expression; few Neopanths except the hardcore mystics accept all of them in this uncompromising form. Many New-Age theorists subscribe to rather hazy or dilute forms of them, whilst some Neo-Pagans have sought to create fundamentalisms all of their own.

1) Nothing is True, Everything is Permitted

This phrase of course intentionally contradicts itself in multiple ways, to create some amusing paradoxes. We could equally well express the implied meaning as;

Everything is True, but only for a given value of Truth.

This does not reflect contempt for reason; rather it reflects an intuition that all truths remain provisional and context dependent.

When it comes to choice of extant religions, Neopantheists often find some sympathy for elements of Hinduism, Paganism, Shamanism and certain forms Mahayana Buddhism. Mainly because they can find plenty of useful symbolism, a wealth of psychological and physiological techniques and a flexible attitude to dogma and paradigm within all of these, despite some of the unpleasant customs in the cultures in which they arose.

Neopantheists usually hold contemptuous views of the three Abrahamic monotheisms. They regard anything that defines itself as absolutely true as obviously false.

If they do have an interest in the abrahamic traditions it usually comes down to looking for allegorical, metaphorical, or heretical material in Kabbala, the Essene mysticism, Gnosticism, and the suppressed gospels and apocrypha.

A similar attitude pertains to science. The best scientific thought always remains provisional and open to improvement or falsification, the worst easily descends to dogma and an absolutism all of its own. Science can only ever make things possible; it cannot in principle prove the impossibility of anything. Neopantheists tend to look upon science as a source of possibility, validation waiting to happen, and ideas often worth borrowing

2) Belief and Intent create Reality

This simple phrase reveals the one and only 'Secret' of magic, mysticism, and all varieties of 'positive thinking'. It's not absolutely true of course. We inhabit a random universe and we cannot always make all of it do exactly what we like. However it works so astonishingly well for much of the time that only fools ignore it. If you don't believe this, then try negative thinking for a while and see where that gets you.

Of course it takes courage and imagination and discipline to develop the beliefs and intents to change a situation, but of all these, imagination needs enticement and encouragement first in the quest for personal empowerment. Thus whilst Neopantheists recognise belief as a tool rather than as an end in itself (faith) they may nevertheless select beliefs which appeal to their imagination and stimulate it further, ritualistically acting out the belief 'as if' true.

3) Alchemy

Nobody believes in Alchemy these days, or do they?

Medieval alchemists seem to have had a variety of agendas. Some simply sought to make gold from other metals and generally failed because they could not concentrate enough energy on their starting materials, although they did discover much about metallurgy and chemistry in the process. Others sought transmutation in a more esoteric sense and tried to turn their own base natures into spiritual gold, they seem to have obtained mixed results although many of them discovered the importance of the Chymical Marriage, the inclusion of the feminine perspective, and worked with a Sorror Mystica, a mystical sister or wife.

Many other alchemists sought medicinal objectives from increased vitality to immortality. Some accidentally achieved quite the reverse effect with heavy metal poisoning, but others seem to have discovered the astonishing effects of what we now recognise as placebo or intent based medicine. The apparent absence of anything materially effective to the scientific view in alternative medicine treatments does not discourage Neopantheists. They delight in the principle of intent and devise analogical or immaterial theories of their own to bolster belief. As you might expect, alternative health practices often fail to perform well in scientifically controlled situations. They need to function as a package on their own terms, snake bones, crystals and all, if necessary.

When conventional medicine administers placebos with full medical ritual the results frequently show better outcomes than those of 'actual' treatments, particularly with medication.

4 The Female Perspective

It seems presumptuous for a male to attempt to define what the female perspective consists of. Nevertheless neopantheism values intuition as much as logic, dreams as much as waking thoughts, psychic experience as much as rational analysis, empathy and compassion and as much as disinterested objectivity, the goddess archetype as much as the god. The neopantheist rejection of the logocentric fundamentalisms with their male monotheist deities and their almost invariably male priesthoods mirrors its sympathy for the female perspective.

5 Synchronicity and Meaning

Neopantheists rely on their personal experiential definitions of reality rather than subscribe to societally sanctioned opinion about what constitutes reality and what doesn't. Thus if a superstition gives good results it gets reused, and coincidence rarely gets dismissed as mere coincidence. We spend most of our lives trying to engineer coincidence between intent and actuality. So if a synchronicity appears spontaneously we should consider interpreting it as an affirmation of deep intent, or a warning from the subconscious. Such 'magical thinking' often attracts the derision of scientifically schooled minds, but magical thinking often produces excellent results when you have exhausted the possibilities of common sense.

6 Sky Fairies or Psi Fairies?

Do gods, demons, spirits, elementals, and discarnate intelligences actually exist?

Well, YES and NO, and YES again, to most Neopantheists.

YES, in the psychological sense that people's gods and demons often do much of the talking in social interaction anyway. So they can pass from person to person.

So we manufacture such phenomena, but they also manufacture us. As biological and social and partially psychic organisms, we consist of bits and pieces from all over.

NO, panpsychism recognises that every phenomenon has consciousness to some degree from the simple consciousness of an atom to the complex consciousness of a brain, but as consciousness consists of a property of material phenomena then it cannot exist in entirely discarnate form.

YES, in the sense that parapsychology and quantum connections allow consciousnesses to effect each other across space and time. Thus in a sense the laws of nature comprise simple and powerful discarnate spirits. Thoughts can act as discarnate spirits also, but generally with less ubiquitous effect.

Sky-fairies in the logos sense exist only inside people's heads, but Psi-fairies, projected from one consciousness to another can create effects analogous to spirits in the classical sense.

7 Personal Narrative and Mythos

Ask most modern westernised people about themselves and they usually reply by describing what they do in terms of profession and interests. They usually lack metaphors for their self or selves although some will reply with some expression of a basic inner metaphor, like I'm a Christian or I'm a Capricorn.

Neopantheists on the other hand prefer an elaborate and extensive personal narrative and mythos. For example, Mercury conjunct with Pluto in Taurus, a Crow as Clan Animal, several half remembered Past Lives, a Spirit Guide, four servitors, a mission to rediscover Atlantean wisdom, and a range of possible future incarnations in mind, plus at least another six impossible things before breakfast.

All this doubtless seems quite deranged to the logocentric mind, but neopantheists would reply that if you are going to have an inner life then you may as well have a large and flexible one and an extensive vocabulary to explore it with.

Who would choose a prosaic inner life, when they could live one of poetry instead?

Magical Thinking of course qualifies you as 'mad' in terms of our current orthodox cultural paradigm. However it merely qualifies you as 'technically inept' if you cannot make it work, within the neopantheist paradigm.

8 Cosmic Holism and Transcendence

Does the universe as a whole; exhibit any kind of consciousness that we can interact with?

Does the universe seek to evolve greater complexity and more sophisticated consciousnesses?

Could it use some help from us in this?

Do all species seem worth preserving regardless of their economic value to us?

Does some mysterious circularity in time connect consciousness and the very existence of the universe?

Most Neopantheists like to think so.

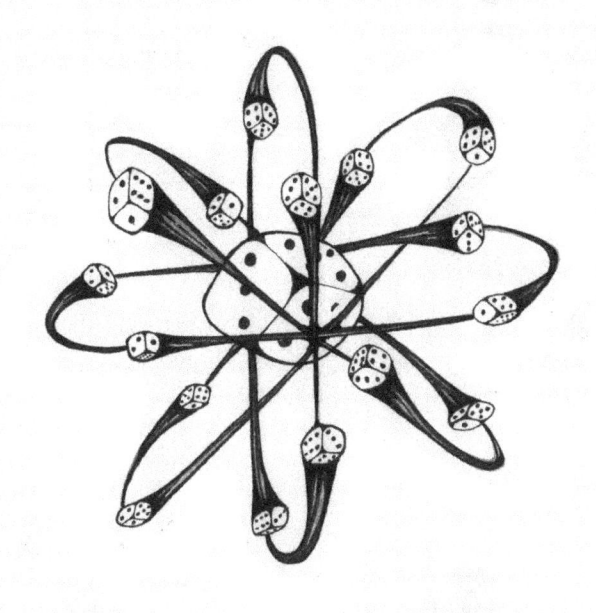

*Dice worlds,
Fractal self-similarity
From Quantum to Cosmos**

$$\Delta J_i \, \Delta J_j \geq \frac{\hbar}{2} \left| \langle J_k \rangle \right| *$$
$(i^2 = j^2 = k^2 = ijk = -1)$

*Indeterminacy in the orthogonal components of angular momentum.
She does spin dice!*

Chapter 5
Metadynamics - Practical Magic

This chapter questions the assumptions of causality and of one dimensional unidirectional time. It examines both the apparent causality failure and the apparent operation of hyper-natural forms of causality implied by quantum physics.

It seeks an Apophenia in a model of three-dimensional time that can model both quantum physics and magic.

> 'It is my opinion that our present picture of physical reality, particularly in relation to the nature of time, is due for a shake up - even greater, perhaps, than that which has already been provided by present - day relativity and quantum mechanics.'
> - Professor Sir Roger Penrose[11]

Part 1. Quantum Weirdness

Quantum physics works beautifully in the sense that it allows us to build all sorts of amusing electronic devices and to model the behaviour of atoms and subatomic particles to a very high degree of precision. However nobody really understands it. The maths gives excellent results, but it contains things like imaginary numbers which have no obvious perceptual meaning in the human scale world. Bizarrely contra-intuitive events seem to underlie the behaviour of the stuff of the universe. Objects can seem to have had several different locations or mutually exclusive states at the same time. Moreover some of the behaviour of quantum entities seems completely random and to arise without prior cause.

Thus many interpretations of quantum physics abound. Some interpretations claim that no underlying reality exists;[12] we have reached down to the simplest level of reality and we just have to accept the strangeness we find there on its own terms. Others seek to find some kind of hidden variable to restore some sort of causality to the apparent randomness of the quantum domain.

Herewith some examples of quantum behaviour to illustrate the weirdness that underlies our reality.

Because our whole language and thought structure revolves around the idea of cause and effect we have difficulty in accepting the idea of random events, and prefer to think in terms of uncertainty instead. We tend to assume that apparently random events must have underlying causes even if we cannot work them out. However nature provides a simple example of uncaused events in radioactive decay.

Radioactive isotopes, (atoms which spontaneously decay), all exhibit a characteristic half life. Plutonium238 has a half-life of 88 years, Tritium (Hydrogen3) has a 12-year half-life, and these half-lives limit the lifespan of nuclear warheads. Many of the Uranium isotopes have half lives of hundreds of millions of years which means that we can still dig the stuff up because some still remains from the formation of this planet's material in an exploding star core billions of years ago. Now a half life denotes the time it takes for one half of a sample to decay, So after 12 years, half of a sample of Tritium will have decayed, after 24 years only a quarter will remain, and after 36 years only an eighth will remain and so on.

Thus the process seems predictable enough, however it seems impossible to explain how this happens except by assuming that each individual Tritium atom has an exactly 50:50 chance of decaying in a 12 year period. The behaviour of the individual atoms would appear to have to remain random, within limits, to produce the half-life effect. Random behaviour means no causal connection to previous behaviour. Just because a dice comes up with five twice in a row does not make it more likely to come up a third time. If a Tritium atom failed to decay in a 12 year period it does not affect the likelihood of it decaying in the next 12 year period; that chance remains 50:50. Dice may not actually exhibit truly random behaviour unless you bounce them around a lot, they may merely exhibit unpredictable behaviour because we cannot calculate all the micro-factors determining how they fall. Nevertheless with the internal behaviour of atoms it seems inconceivable that some sort of internal micro-factors generate the observed behaviour. Quantum physics depends on the idea that nature does not have unlimited divisibility, at some point something comprises the smallest possible piece of reality. It won't have any internal structure or smaller components within, and at that point the chain of cause and effect must presumably come to a halt.

The Double Slit experiment provides a second example of the weirdness of quantum behaviour. This seminal experiment demonstrates the whole mystery. Many variants on the original experiment exist but they merely serve to confirm the mystery a little.

If you fire light quanta or electrons or even moderately large molecules like Buckyballs (consisting of 60 carbon atoms), at a screen with a small hole in it, then they pass through the hole and land on a target the other side as you would expect particle like projectiles to behave. If you use a screen with two holes in it then they land on the target in a particular pattern as if as if they had passed through the holes as waves instead, even though they land on the target as particles. The wave like aspect of their behaviour suggests that they do not have a definite location in space and time whilst in flight, but that they somehow smear themselves out over a range of spacetime locations. When they encounter a target they somehow collapse back into definite particles, but their wavelike flight mode allows them to do seemingly impossible things.

All objects have wavelike characteristics, but things as large as bullets have a wave function much smaller than the size of a bullet, so bullets tend to go through only one of two closely spaced holes in a steel plate. However tiny objects like light 'particles', electrons, and moderately large molecules, seem to have the ability to pass through both holes simultaneously because their wave functions have a similar size to their particle sizes.

We should not however suppose that the wave like characteristics of quantum entities limits the weirdness to tiny areas of space much smaller than human scale events. With the progress of time, the wave functions can become spatially huge. Instead of using a screen with two closely spaced slits in it, you can use a half-silvered mirror to give a beam of light a choice of directions in which to proceed. Light quanta can either go through it or reflect off it, and with this you can achieve quantum weirdness on any scale you like. It seems that with such a 'beam splitting' apparatus we can force individual light particles (for this is how they manifest at the detectors) to fly 'both' ways round a system of mirrors that we can position yards or even miles apart. The wave function can become enormous by human standards. At this point it becomes imperative to take care about 'when' we speak of. Before a particle sets off, it may appear to have a choice of trajectories, when it lands it may appear to have exercised both choices simultaneously, we cannot however investigate its apparently

67

wavelike manifestation whilst it flies, for in doing so we force it to collapse back into particle mode.

That a half-silvered mirror can apparently split a single light particle into two waves says something utterly strange in itself. Light registers on detectors by getting absorbed by single atoms in the detectors, yet a half silvered mirror consists of little clumps of silver atoms that reflect light particles instead of absorbing them, and spaces between the clumps where they can pass through. So although individual atoms can absorb light particles they appear to have a fairly huge wave size compared to an atom whilst in flight because even a fairly coarse grained half silvered mirror that looks patchy under a hand lens will do the trick.

The presentation of electrons that you get in elementary chemistry and physics classes as tiny little electrically charged balls orbiting the nuclei of atoms or travelling down wires to supply electrical current gives a model of very limited explanatory power. For chemistry to work as we observe it, the electrons need to act as though they have a sort of smeared out existence all over the outside of the nucleus. They don't function as tiny little balls whilst in orbit, they act like diffuse spherical clouds englobing the nucleus, but in other situations they act as point particles of zero size.

At the quantum level particles seem to behave as if they can 'be' in several different states at once or 'be' in several different locations at once. However we can never observe them in such a condition, we can only make observations that strongly suggest that they had occupied such states prior to our measurements. Here we see the double slit mystery re-appearing. Single particles appear to have passed through two different states simultaneously. This phenomenon has the name of superposition and it dominates the way the universe works. Most of the particles of mass and energy that make up the universe seem to spend most of their time in superposed states. Only when they interact with each other do they seem to fall out of their superposed condition and momentarily manifest in a definite particle like state. The collapse of the superposed wave state occurs randomly, but because most human sized events involve billions of particles, such behaviour creates a more or less perfect illusion of cause and effect, at least in the short term. Thus whilst the water molecules in the glass on my desk vibrate and jiggle around quite violently and keep dropping into and out of superposed states, the water as a whole keeps fairly still and its behaviour remains fairly predictable. Yet some individual molecules may occasionally escape the surface of the liquid and evaporate away.

Under certain circumstances the collapse of the wave function of particles occurs in a not entirely random way, this happens if the wave functions of two or more particles become entangled. Quantum entanglement seems to contradict all the normal assumptions that we acquire about causality, space, and time. Many variations of the basic entanglement experiments exist, but a generalised account of what happens goes like this: Allow two particles which have come into contact to travel off in different directions, then force one of them to collapse its superposed state and assume a definite particle like property. You can choose what property to measure but randomness ensures that the answer will come out as either yes or no for that property. Now in doing this you ensure that the other particle will give a no if you got a yes, and a yes if you got a no, and this seems to work across any amount of space and time you like. Thus not only do particles spend most of their lives in superposed states, but those superposed states remain entangled with those of the last thing they collided with. So if your eye caches sight of a distant star at night it establishes a quantum connection to an event billions of miles and perhaps thousands of years ago.

Conversely, and here it gets really bizarre, as you look out at that far star at night, light from you can in principle entangle you with an alien not yet born, thousands of years in the future, on a planet orbiting the faraway star.

With reality appearing to behave so differently at the quantum level than it appears to behave on the macroscopic level, many people have sought to interpret quantum physics in a way that makes some kind of sense in macroscopic terms. Often this has meant trying to add some kind of hidden variable to sneak causality back in, but none seems convincing. Macroscopic events do however differ from quantum scale events in one important respect; they exhibit a preference for increasing entropy. Processes involving huge numbers of particles do not usually exhibit time reversibility. Eggs break fairly easily but broken eggs never seem to unbreak, and a time-reversed film of an egg reassembling itself from broken pieces looks unrealistic.

On the quantum scale, events seem less limited by this apparent one way restriction in the direction of time, and the equations describing many quantum changes look fully reversible in their relativistic form, so nothing seems to prevent them happening in reverse.

So, in summary quantum physics presents us with two phenomena to reconcile with the rest of our understanding of the universe, namely superposition and entanglement. Both of these seem more comprehensible if we assume that what we observe as particles actually have a wave like behaviour that spreads out in both space and time into the past and future of the moment of observation. After all, superposition implies hyper-temporality, superimposed events happening at the same time, whilst entanglement implies hyper-locality, linked events happening at the same time in different places.

One particular interpretation of quantum physics, Cramer's Transactional Interpretation,[14] explicitly describes the double slit experiment in terms of phenomena moving both forward and backward in time. In this model a forward wave goes through both slits and then makes the target emit a time-reversed wave, which travels back down one of the two paths at random, taken by the forward wave. The time-reversed wave meets the forward wave at every point of its trajectory and the two waves combine to make a particle. Thus in a sense, the particle reality arises out of an overlap between waves coming from the past and the future. This transactional scheme also makes some sense of the phenomena of superposition and entanglement. We can never observe superposition actually happening because any attempt to observe it forces it to collapse. Nevertheless it often seems that we observe behaviour that could only have arisen from a superposed state. Now if the past of a particle consists not of a discrete single state, but of two or more waves, then the moment of a particles interaction or measurement marks the point where these waves overlap and collapse to create a particle- like effect.

Similarly in entanglement we do not need to posit some incredible action at a distance that somehow finds its precise target across vast tracts of space and/or time. We just need time reversibility. When one of a pair of entangled particles falls out of superposition it sends a time reversed wave back down its trajectory back to the point where both particles had contact. This then modifies the starting conditions, which in turn ensures that the other particle in the entangled pair behaves appropriately.

Time reversibility thus solves the problem of how a single particle can 'know' that a screen has two slits, and how it can 'know' what it's entangled partner has done on the other side of the universe. However it does not explain the randomness or the apparent superposition of two states in the same 'place'.

For this I suspect that we need not merely reversible time but three-dimensional time as well, time which extends 'sideways' as well as just fore and aft. I propose that time may thus have the same dimensionality as space, three in each case. This may seem rather contra-intuitive on first analysis, after all a calendar shows a string of dates in a row but it never shows extra days stretching out sideways from any day, nor do we seem to experience such things. We do however generally accept that a number of possible tomorrows might follow today, although most people seem to assume that a singular yesterday led to today, despite that historians argue interminably about how and why we arrived at today. The assumption of a singular past will receive some re-examination in the following section.

Part 2. Three-dimensional time

If time does have a three-dimensional solidity we would not see it directly. We cannot even see a fraction of any length into the past or future by normal means anyway, so a thickness in time would generally go unnoticed as well. However a universe with sideways time would have one defining characteristic in particular; it would appear to run on probability rather than on strictly causal deterministic principles, and this one does.

Time appears linear and one-dimensional because we define and measure time as the direction in which entropy increases, but entropy only appears on the macroscopic scale, where large numbers of particles participate in a process. Although various macroscopic processes lead to increasing entropy at different rates we have tended to adopt the revolution of heavenly bodies as our standard entropy-meters as they dissipate their energy only extremely slowly and at a fairly constant rate.

Probability lies at right angles to time as we measure it, in sideways time, and it acts as a sort of pseudo-space or parallel universe space, but we should not suppose that any of the 3 dimensions of time has a special status, anymore than any of the spatial dimensions has. Now all objects have a limited spatial displacement in three dimensions, two and one dimensional objects exist only as theoretical idealisations; a piece of paper must have some thickness to exist. Similarly all objects have a displacement in 3 dimensions of time as well. Their temporal 'thickness' at any instant equates to their wave property, and it has enough room to accommodate superposed states which have slightly different orthogonal time coordinates. Thus at any instant of the present not much temporal room exists for parallel universes because particles displace only tiny amounts of time. Most of the particles in my body will exist in superposed states at any instant, but that does not imply that overall I exist in many parallel universes in any meaningful way. My overall wave property at any instant does not much exceed that of the size of a single particle. Thus it serves to locate me fairly precisely in time and space on the macroscopic level, even though most of the particles inside me have multiple orthogonal time coordinates in the pseudo-space of parallel universes.

Noether's theorem asserts that all conservation laws reflect symmetries in nature in which something remains constant. Thus for example the claim that 'matter can never get created or destroyed' implies that the amount of it remains constant under time translation. This claim proved inaccurate, and Einstein replaced it with the celebrated mass-energy equivalence where the energy equals the mass times lightspeed squared. This new conservation law asserts that the total mass-energy remains constant in time although one can change into the other. Heat an object and it becomes heavier, but only infinitesimally so at kitchen temperatures.

Einstein also uncovered a non-obvious space-time equivalence. All objects always move at exactly the same rate in spacetime, despite appearances to the contrary. The faster something moves through space the slower it moves through time. Onboard time actually slows down for objects moving very fast, months of jet travel can take a few fractions of a second off an accurate clock and theoretically add them to the life span of those travelling with it.

We measure time only by movement in space, even if that movement consists merely of parts moving within a clock or within the human body. A deep symmetry exists between space and time, so why do we ascribe different dimensionalities to them?

Large pieces of matter each move only in one direction in space at a time on the macroscopic scale, thus we need only one dimension of time to describe their motion to a reasonable approximation. However if something did move in several directions in space at once then we could use a three-dimensional time frame to describe it.

Can anything actually do this?

Yes, the wave aspects of particles of matter do it all the time, but usually on such a small scale that we do not notice it, in the same way that we do not usually notice the mass-energy equivalence or time dilation at speed. However waves sometimes have very big effects which show up as quantum entanglement over many kilometres or in the capricious phenomena of magic.

When it comes to the past and the future, objects can have as much orthogonal time as the period of 'ordinary' time under consideration, this equates to the idea that events become progressively less predictable or determinate the further you look in time. So a particle has many possible futures and its wave like behaviour allows it to spread out and 'try' all of them to some extent, but it only gets feedback across time from one possible future at random. This then creates positive interference and allows the particle to manifest in some definite form in the future.

Despite that we assume the past to exist in singular form because we experience our own past singularly, both magic and quantum physics suggest otherwise.

From the standpoint of the present, the past and the future do not exist in definite form. The present consists of the moment of interaction between waves from the past and the future as they collapse randomly into particle mode. The past and future consist entirely of wave modes spread out in orthogonal time in a progressively more diffuse fashion the further you consider them from the present. Thus time travel into the past remains a silly idea because the past merely consists of wave like echoes of what might have been. Time travel into the future remains possible, but only if you isolate yourself from the effects of entropy by slowing down your onboard time by travelling at, or accelerating towards, something close to light speed.

Nevertheless in both magic and quantum physics you can modify what probably happened in the past, so long as it does not alter the present, and you can see that you have done this because the future then manifests in unexpected ways. Magical literature abounds with anecdotes which strongly suggest that some enchantments have their effect by modifying the past, and the Delayed Choice Quantum Eraser version of the Double Slit Experiment demonstrates this effect convincingly enough. In this experiment a subtle arrangement of devices allows you to choose whether or not to preserve an observation of which slit a particle probably went through, and such a choice then seems to actually modify whether it 'did' or not.

I have called the 3-dimensional reversible time interpretation of quantum physics 'General Metadynamics'. Like most of the other interpretations it remains un-falsifiable at the time of writing, and thus to a certain extent it remains a matter of taste. However two related lines of speculation do lend support to the idea of 3-dimensional time.

Firstly the structure of the suite of currently known particles of matter and energy does supply an unexpected source of possible confirmation.

Appendix (i) deals more fully with the technical side of this argument, but in brief; three varieties of all fundamental matter particles have been found. The ordinary ones make up the overwhelming majority of the stuff of the universe, but two heavier versions of each exist. These heavier versions rarely appear in nature but we can make them, although they have short lifetimes. The number three seems to dominate particle properties. Strong nuclear charge occurs in 3 varieties, electroweak charge also manifests as a fraction or whole of three basic units. Appendix (i) shows how the extra degrees of freedom afforded by three dimensions of time allow particles to have spins which account for these phenomena. In particular the hypothesis explains why the two heavier and apparently superfluous extra versions of matter particles have to exist, and why charges manifest in threes. Of course the reversibility of time also leads to corresponding anti-charges and anti-particles, again in groups of three, which we can observe.

Secondly, if the universe exists as a finite and unbounded structure in space and time then it probably has the geometry and topology of a vorticitating hypersphere which will mean three dimensions of time as well as three of space. Chapter 6 and its appendices attempt to clarify this heretical idea. Yet for now I'd like to examine the magical implications of the general metadynamics interpretation.

Part 3. General Metadynamics and magic

A few months' examination of a library of magical books might well give the impression that the whole subject appears abominably complex and impossible to reduce to any sort of comprehensible structure. However if we ignore for a moment the mythos and symbolism and metaphysical paradigms adopted by various traditions of magic and concentrate instead upon the actual objectives sought, and techniques used, then it all begins to look a good deal simpler.

The basic ideas of magic, which have remained with humanity since the dawn of thought, and which the earliest traditions of shamanism seem to have preserved, reduce to five core ideas: -

1) Divination. The idea that certain practices can reveal information by non-ordinary means.

2) Enchantment. The idea that certain practices may encourage desired events to occur by non-ordinary means.

3) Evocation. The idea that by certain practices people can command 'spirits' to assist with divination or enchantment objectives.

4) Invocation. The idea that by certain practices people can enter into some sort of identification with, or possession by, 'spirits' to achieve divination or enchantment objectives.

5) Illumination. The idea that certain practices enable people to gain special knowledge and powers that ultimately seem to reduce to divination or enchantment.

Thus divination and enchantment remain the basic measure of magic because we know enough about the mechanisms of evocation, invocation and illumination by now to understand that these practices act as psychological mechanisms to support attempts at divination and enchantment.

Debate of course rages about the 'certain practices' that give the best results in each of these five activities. These 'certain practices' actually remain rather uncertain and somewhat ad hoc and rule of thumb at the time of writing. However the hypothesis of physiological 'Gnosis'[16] and the hypothesis that 'Sleight of Mind'[17] can unleash the subconscious, have helped to refine the practices towards something approaching a reliable toolbox.

Divination and enchantment constitute the core of what some have called parapsychology. This word has perhaps less usefulness than it seems, because if its effects exist, then it implies something more general about the universe that goes beyond mere psychology to imply a whole Para-Physics which begins in the quantum domain and protrudes capriciously into macroscopic reality as magic.

The General Metadynamics interpretation of quantum physics provides a paradigm that can model the divination and enchantment effects underlying what we call parapsychology if we add the concept of Decoherence.

Decoherence explains why quantum effects do not dominate the macroscopic world. A photon lucky enough to fly from Sirius to your eye without hitting anything along the way can remain in entanglement with the electron that emitted it on Sirius a decade or so ago. This can happen mainly because few particles get in its way in the intervening space.

On the other hand Schrödinger's hypothetical cat, whose fate depends on whether or not a quantum event triggers its death inside a sealed box, almost certainly exists at all times in either a dead or an alive state inside the box, irrespective of our observations or lack of observations. This happens because entanglements rapidly get out of phase as particles interact with other particles in their environment. The 'yes/no' wave state of the particle controlling the cat's fate cannot entangle coherently with the entire cat and put it into a state of 'life/death' superposition because as particles interact the coherency rapidly becomes lost amongst the jumble of atoms comprising the apparatus. Thus the cat killing mechanism as a whole remains either triggered or un-triggered, the superposition of the quantum state controlling the mechanism fails to entangle coherently with much of the mechanism. However at some randomly chosen time when the superposition does collapse, the mechanism does one thing or the other, although we cannot predict when it will do so.

Superposed and entangled states exhibit great delicacy, they remain very prone to decohering into their environments by contact with surrounding particles and this has raised a serious barrier to the construction of quantum computers. A quantum computer can in principle explore a vast number of possible answers to a question simultaneously by using components that can apparently pass through many superposed alternative states at the same time, however the critical quantum parts of the components require very careful isolation from their environment to prevent decoherence.

The brain functions as a rather chaotic analogue computer. A given input to the brain or even to a single of its component neurones, does not always elicit the same response or the same strength of response. Relatively long range connections between different parts of the brain tend to work rather erratically, and this leads to more unpredictable function. Much of the brain seems to function on threshold effects rather than simple digital on/off type effects. As a stimulus strengthens, the probability of a response increases, but its effect remains unpredictable at lower intensity. Sometimes butterfly type effects occur; a single idea can initiate a mental cataclysm. At the time of writing we have very little idea of how the brain stores memory, although we have a rough map of where it seems to store it. Curiously it seems to store memory in the same areas that it uses to imagine and anticipate the future.

Magic works in Practice, but not yet in Theory.

Well it may not work very reliably in practice but the balance of evidence from parapsychology does suggest that it does play a limited but real role in reality. Divination and enchantment do sometimes achieve statistically impossible results.

The theory however remains problematical. If we choose to abandon the antique hypotheses of spirits, transcendental agencies, and mysterious aethers, then only quantum ideas remain as possible models. In this case the brain must somehow allow some quantum effects to manifest at the macroscopic level.

The brain must operate not only as a chaotic analogue computer, but to some extent as a Chaotic Analogue Quantum Computer as well.

A Chaotic Analogue Quantum Computer might sound like a rather crazy specification for a brain but it accords rather well with our subjective experience of 'mind', the activity which the device performs.

States of Voidness can arise from either mind stilling meditation or in milder form by absent minded distractions. In this condition the brain seems to relax parts of itself into states of Superposition pregnant with possibilities out of which inspiration can collapse. Sometimes divinatory phenomena manifest in these states.

States of Gnosis can arise through the physiological experience of extreme excitation or extreme focus of the nervous system and this seems to correspond to Coherence, with extensive areas of the brain all exhibiting the same mind activity whilst the function of other areas becomes strongly inhibited.

It seems that the brain may have the ability to somehow preserve superposed states so that they can remain Entangled with past and future brain states. Divination thus works because the diviners basically have access to some future state of their own brain when it knows the answer. Divination experiments in which the diviners themselves will never know whether they divined accurately or not, usually fail abysmally. The tendency for superpositions and entanglements to decay over time would then supply an additional reason for divinations to tend to work best for short time periods. The great majority of my premonitionary experience tends to occur just a few minutes before the event.

Some magicians make a point of trying to visualise themselves at a future time when they will have found out the answer to a divination. They may also resolve to visualise sending back the information to their current divining self when they have it, to establish a closed loop in time.

Entanglement of present brain states with past and future ones can also provide a model to explain enchantment. Enchanting into the future presents the simplest case.

If by techniques such as Visualisation coupled with Gnosis the magician can establish a future brain state which perceives a desired event as having come about, then physical reality will have a tendency to decohere towards a situation in which it has.

This strongly suggests that when enchanting for a future event, magicians should focus on establishing a future perception or 'memory' of it having occurred, rather than visualising a chain of events leading to its occurrence.

Thus 'On my fortieth birthday I have magnificent property assets', makes a better statement or visualisation of intent for a spell than 'It is my will to become rich by the age of forty'. The former spell encourages the whole of entangled reality to work towards your desire, whilst the latter merely increases the chances that you might make the right choices.

Retroactive enchantment appears to work by a similar mechanism. 'At twenty three I have a series of life changing experiences which equip me well for the future'. Such a spell might usefully undo many of the negative effects which seemed to stem from the experiences at the time, both on the psychological and physical levels. A retroactive enchantment cannot take place if it alters the measurable conditions of the immediate present and thus prevents itself occurring, so we can only measure its effects by the amount that it causes the future to deviate from its probable course.

Quantum Entanglement underlies the idea of the magical link and antique theories of magic by contagion, yet it sets rather severe limits on what we can achieve with it because decoherence tends to weaken the effects of entanglement. Simultaneous physical presence with physical or line of sight contact seems to offer the best chance.

Artefacts once connected to the target or visualised remembered images come in at second best, whilst photographic images qualify as a rather poor third choice, real time live images or telephone calls may offer better possibilities if you can establish them.

General Metadynamics, the quantum-magical hypothesis of three-dimensional reversible time, has its own equation:-

$$\Delta S K° \Delta t_3 \sim \hbar$$

This represents a new member of a class of equations called 'Uncertainty Relationships' that follow on from Heisenberg's celebrated equation relating the uncertainty (and almost certainly the actual indeterminacy) of position and momentum.

It means that the indeterminacy in the entropy S, times the indeterminacy in the time t, (in any of its 3 dimensions), has about the same magnitude as Planck's constant, h. (Note that we need to specify the absolute temperature, K, (at which we measure the entropy, to preserve dimensional equivalence, but this makes little practical difference).

Thus any activity of the universe which constitutes a minimal entropy change can proceed for plenty of time. So a particle can 'feel out' multiple possible future trajectories so long as only one of them gets reinforced by reverse time feedback to become real, as the options it didn't actualise create only infinitesimal entropy.

Thus we can think of time in three dimensions as working by a process of Apophusis, Apophasis, and Apoptosis. These Greek derived words have acquired various applications in biology, rhetoric, and biology respectively, but they illustrate the underlying mechanisms of reality:

Apophusis - branching, reality makes a feint at every possible thing it could do.

Apophasis - weirdness, what doesn't happen may still have an effect on what does.

Apoptosis - dying off, a collapse of superposition and entanglement to yield a result.

Curiously, at least on a subjective level, the mind feels that it works like this as it seeks decision, inspiration, or Apophenia. This suggests some sort of quantum-panpsychic principle at work in both the microcosm and the macrocosm.

Part 4. On The Nature of Time

> What then is time? If no one asks me, I know what it is. If I wish to explain it, I do not know.
> - Saint Augustine.

The present seems to exist for a fleeting instant only, the past seems to exist in memory only, and the future seems to exist in our expectation only.

(Note that all the records cosmological, geological, literary and in the form of memory, exist in the fleeting now, and structure our beliefs about the probable past and the possible future).

Does time exist? Can we ask what it 'is'? Do we perceive time or do we construct it as a working hypothesis?

I have a device that shifts the entire universe lock, stock and barrel, every last particle, a million years into the future (or the past) every time I activate it, but nobody ever notices.

Only a record of relative movement and change seems to give us a sense of time. Plainly time does not exist as something abstract and separate from movement and change. Time does not flow and it has no location.

I submit that we have difficulties in forming a coherent picture of time because the past and the future consist of something radically different from the present.

The universe consists of quanta that sometimes appear to behave as particles and sometimes appear to have behaved as waves. Note the careful wording here, we can never catch a quantum behaving as a wave, we can only catch it as a particle. After we have caught it we can say that it appears to have behaved as a wave to arrive in the position we caught it in. Similarly for the future we can only make a prediction about its wave behaviour and the range of possible particle states that might lead to.

For large lumps of matter we can usually ignore the wave behaviour of the constituent quanta because the wave behaviours tend to cancel out and allow us to establish fairly reliable memories and expectations. Thus we can construct working hypotheses of cause and effect, and get away with the idea that the past and future have a similar reality to the present moment.

But of course they do not; we create that illusion by memory and expectation and with ideas about cause and effect.

The present moment always manifests in the singular as a particle-like reality. The past and the future of any moment of the present have a wave-like reality.

The past and the future consist of a vast array of waves forming a much 'larger' universe than the one we observe directly, it forms a multiverse of wavelike parallel universes out of which the observable singular particle-like universe of the present moment appears as an interference pattern. This occurs as a two way process, the particle-like present subtends the wave pattern into the past and future multiverse but the multiverse also subtends an interference pattern to create a fleeting particle reality.

This can only happen because time has three dimensions, it has 'width' to accommodate all possible pasts and futures, not just the length in which to accommodate a single past and future.

The whole idea of 'being' thus seems illusory and to merely arise from our rather sluggish perception which fails to notice the ubiquity of change.

The whole idea of the past and the future thus also seems illusory because no particle-like reality exists there at all.

We learn to conjure an illusory picture of reality for ourselves in which we, and other people, and various phenomena have 'being' and some sort of a 'real' past and future, from the perspective of the present. Without that illusion we would probably find existence intolerable.

The above paradigm represents General Metadynamics taken to its logical conclusion.

It provides a model of the physical principles underlying both quantum physics and magic.

Yet I regard it as a dark illumination, an unpleasant insight into how the machinery of the universe may actually work, I find it at least as disturbing as the idea of the inevitability of personal death.

Yet as a Chaoist I must regard nothing as true, but regard some things as having greater or lesser degrees of utility.

Thus I will use it for magic as I find it the most convincing paradigm available, despite that I find it mystically unattractive. For the purposes of conducting my ordinary life I shall use other less austere paradigms.

Ouroboros,
An Alchemical symbol representing
a subtle blasphemy;
The finite and unbounded curvature
Of the eightfold universe,
Moreover, it lives…

Chapter 6
Non-Singularity-Cosmology

Introduction. This chapter seeks to undermine the notion that the universe must contain some kind of ontological singularity or metaphysical catastrophe like an infinity, or a Big Bang beginning, or a Big Crunch ending, or a god to start and finish it. Such things put its existence beyond rational understanding in principle because they introduce a profound self-inconsistency, the physics on which the universe runs breaks down at a space-time singularity and god based explanations usually supply nothing more than an excuse to selectively abandon rational enquiry altogether.

This chapter seeks an Apophenia in the idea that any real quantity has a finite yet sometimes unbounded extent, and that no real quantity can have an infinite value.

Thus it attempts to undermine the whole linear time paradigm of occidental and monotheistic thought which endows the universe with a beginning and perhaps an end.

Instead it posits a universe that consists of a finite and unbounded amount of both space and time and which exists naturally, simply because it exhibits physical and magical self-consistency.

Part 1. Against Singularity

An erroneous consensus has developed amongst astronomers in recent decades that the universe began about 13 billion years ago with some kind of a big bang. Three lines of evidence have led to this conclusion.

Firstly the light and other electromagnetic radiation from distant galaxies has less than the expected amount of energy when it reaches us. As light invariably travels at the constant light-speed in free space, this energy loss appears as a red-shifting of the light towards the lower energy end of the spectrum. Astronomers interpreted this as evidence that the universe had expanded from a much smaller size in the past and that the expansion of the universe had stretched the light waves out, thus increasing their wavelength and lowering their frequency and energy. At first it seemed that the amount of redshift corresponded roughly to distance, implying a constant rate of expansion or perhaps a rate which had slowed slightly over time due to gravity. More recent observations seemed to suggest that the expansion rate had somehow increased with time. As a logical consequence of this hypothesis it seemed that the entire universe must once have had virtually if not actually zero volume and an infinite or near infinite density. Observations of the redshifts of very far galaxies suggested that they recede from us at velocities approaching light-speed.

The second item of evidence comes from the cosmic microwave background radiation or CMBR. A light and uniform drizzle of microwave radiation comes in from all directions in space, indicating that most of it comes from very deep space beyond our galaxy. Astronomers interpreted this radiation as a remnant from the very hot fireball state in which the early universe supposedly existed. By now the expansion of the universe had supposedly cooled the radiation of the primordial fireball down to weak microwaves.

A third item of evidence depends on a circular argument. The universe appears to consist of about 75% hydrogen and 25% helium with just a tiny smattering of the heavier elements. Now from what we know of the synthesis of helium and the heavier elements in stars from hydrogen, the stars have not had long enough to make all the helium we can observe if the universe started with only hydrogen about 13 billion years ago. Thus astronomers concluded that the primordial fireball itself must have made most of the helium.

Now the big bang theory which developed from these interpretations of observations suffers from very many problems which theorists have attempted to overcome with a variety of theoretical patches, fixes and fudge factors which have created even more problems and inconsistencies.

Herewith a small selection of some of the most serious problems:

Nobody has a convincing explanation of how the entire universe could have got into the absurdly unnatural state of zero size and infinite density in the first place, or even how it could have expanded out of this condition.

Nobody has a convincing explanation for the apparent uniformity of the universe on the very large scale; such uniformity does not seem a likely consequence of a big bang. The cosmic inflation theory attempts to solve this problem by supposing that space itself somehow expanded at virtually infinite speed to create a universe of the size we now observe, or possibly a much larger one, and that the matter and energy expansion followed afterwards. No credible mechanism exists to support this hypothesis.

Nobody has a convincing explanation of why our best theories of gravity contradict the big bang hypothesis. Theorists have attempted to tinker with gravity theory and to introduce extra sources of gravity and anti-gravity rather than question the big bang orthodoxy. Few professional theorists have dared to doubt the big bang hypothesis itself. At the time of writing, such a policy looks like a suicidal career option on a par with taking a professional interest in parapsychology.

It appears that many galaxies do not contain enough matter to explain how they manage to rotate at the speeds we observe without flying apart. Conventional theory favours the idea of so called 'dark matter' to balance the maths. This stuff does not consist of anything even remotely like the stuff that comprises this planet, our star, and us, yet according to theory it should comprise a substantial fraction of the entire universe. Its properties imply that we can never actually get hold of a bucketful of the stuff and test the idea.

A minority conventional theory called MOND, modified Newtonian dynamics, merely adds whatever fudge factor you need to balance the equations, without offering a mechanism.

The apparent acceleration of the apparent expansion of the universe has led theorists to posit the existence of so called 'dark energy'. If it exists, such dark energy must comprise the majority of the energy in the universe. Yet it must have the astonishingly convenient ability to exhibit anti-gravity to force the universe to expand in an accelerating fashion, and simultaneously the ability to exhibit ordinary gravity to make space appear geometrically flat.

Such hypothetical substances as dark matter and dark energy begin to resemble the Phlogiston which medieval theorist invoked to explain why things burned. Set a piece of wood alight and you end up with a much lighter pile of ash at the end. Ergo the wood must contain Phlogiston that appears as fire and accounts for the weight loss.

When some bright spark noticed that the residue from burning metals actually weighed more than the original metal, (we now know that burning metals absorb oxygen), the Phlogiston in metals then got credited with negative weight, whatever that means.

Nevertheless, despite the highly dubious patches and fudges required to keep the big bang theory afloat, the majority of professional cosmologists confidently assert as fact the idea that the universe consists of about 10% ordinary matter, 20% dark matter, and 70% dark energy. Their jobs depend on it.

Cosmologists are seldom right, but never in doubt, as the old saying goes.

However a far simpler explanation exists for the observed galactic red shifts, the CMBR, galactic rotation rates, and helium abundance. It does not involve a big bang, or fudge factors like dark matter, arbitrary adjustments to gravity theory, or dark energy, or an unexplained preliminary inflation of the universe, and absurd initial conditions.

It simply suggests that the universe has a small positive space-time curvature and thus that it exists as a finite and unbounded closed structure (a hypersphere) in both space and time which undergoes a very slow kind of special 'rotation' which prevents it from collapsing. Part 2 of this chapter gives a verbal description of such a structure and Part 3 discuses the philosophical, metaphysical and magical implications of this model. The mathematics which describes it precisely appears in Appendices (ii) and (iii).

Part 2. The Hyper-Spherical Universe

If nothing can exceed the speed of light, as special relativity asserts and experiment appears to confirm, then any structure with enough gravity to have an orbital velocity of light-speed will function as a 'closed' region of space-time from which nothing can escape. Anything, including light, which attempts to escape, will simply fall back in again or just keep on going round and round forever. The gravity of the structure basically makes space (and time) curve back in on itself in accordance with the theory of general relativity which describes gravity not as a force, but as a curvature in space and time.

Einstein originally thought that the universe consisted of a structure like this, but he had to add a fudge factor which he called the cosmological constant to stop it collapsing in on itself under its own gravity, because it plainly hadn't done so already.

Gödel came up with the idea that the Einstein universe might rotate and thus not collapse, in the same way that the orbital velocity of a planet stops it plunging into its star. However Gödel's model treated the universe as a sphere which would have had an axis of rotation. This would have showed up fairly obviously to astronomers and it didn't. Then the red shift data appeared and the idea of an explosively expanding universe replaced that of a static universe maintained by a mysterious cosmological constant.

A gravitationally closed universe has a positive space-time curvature and the geometry of a hypersphere. Now a hypersphere represents a higher dimensional version of a sphere in the following way. We can consider an ordinary sphere as a two dimensional surface bent round in a third dimension to create a ball, so that the surface no longer has edges. The simplest hypersphere, the so called 3-sphere, consists of a three dimensional volume bent round a fourth dimension to form a structure which has no edges either, it joins up with itself rather than having edges.

To visualise a hyper-sphere consider the possible ways of making a flat map of the earth, they all involve some kind of distortion, but we will have to distort the hypersphere a bit anyway as our visualisation abilities do not work too well in more than 3 dimensions.

You can cut a globe of the world into two hemispheres across the equator and place them next to each other and take a photograph of them. This creates a so-called polar projection that gives a realistic view of the Arctic and the Antarctic but tends to distort the equatorial regions. In such a polar projection, the two circles showing the northern and southern hemispheres normally get placed in contact at some arbitrary point. This reminds us that the now divided equator actually remains in contact with itself at all points, so we could roll one circle around the other to any position to show this. Using this idea we can form a fairly good mental model of a hypersphere. A hypersphere would consist not of two circles in contact but of two spheres in contact, with the proviso that the spheres are actually in contact at every point on their surfaces, which we can represent by continuously rolling the spheres around over each others surfaces. In such a situation nothing can escape the structure. If anything exits the surface of one sphere it immediately enters the other one at the corresponding point on its surface. The division of the hyper-sphere into 2 spheres does not imply any sort of division in reality or any special status accorded to the centre or surface of either sphere. When we cut the world globe into two hemispheres, we can 'cut' it anywhere for representational purposes. We could cut it across the Greenwich meridian and dateline to show an east and west hemisphere if we wanted.

We do not have to centre such projections on the north and south poles. Similarly the centre points of the two representational 'halves' of the hypersphere have no special status, the hypersphere has no centre in the same way that the surface of the earth has no special centre points.

However a hypersphere has a similar property to the surface of an ordinary sphere in that any point in it has a corresponding antipode point which represents the furthest point that you can travel to from the original point until you start coming back towards it from the other direction.

The above description shows the properties of the three dimensional 'surface' of the hypersphere. Technically speaking a hypersphere exists as a four dimensional structure with its 3 dimensional surface embedded in four dimensions, much as an ordinary sphere consists of a two dimensional surface bent round to achieve closure, embedded in a three dimensional space. The fourth dimension of a hypersphere does not have to extend beyond the three dimensional surface. It can consist merely of the curvature of the three dimensional surface which results in the 3D 'surface' having a slightly higher internal volume than it would appear to have if you could look at it from the outside, and assumed that it consisted of a sphere.

Now a hypersphere has several properties which theorists failed to take into account when they discarded it as a model of the universe in favour of an expanding model.

A hypersphere can have a kind of rotation but this consists of something a little more complicated than the simple rotation of an ordinary sphere about an axis, like the north-south axis of our planet. A hypersphere rotation consists of a rotation of the three dimensional surface volume about the radius of curvature, which lies at right angles to all of the three spatial dimensions. We should more properly call such a rotation a 'vorticitation', we cannot easily visualise it, but it corresponds roughly to the idea of a ball of dough kneading itself. In effect every point in the hypersphere changes place with its antipode point and then returns to its original position to complete a single vorticitation. In a universe of this size it would take about 22 billion years, yet it would create a centrifugal effect which exactly balances the centripetal effect of the gravity or positive spatial curvature of the universe. Thus a vorticitating hypersphere can remain stable without collapsing or having to expand.

The combined effect of the centrifugal and centripetal effects in a vorticitating hypersphere would produce a small resistance to linear motion in any direction within the three dimensional space. We have already observed the deceleration of space probes dispatched some years ago to the extremities of the solar system. This so called Pioneer Anomaly or Anderson acceleration has led to much debate and argument among theorists. However if it does represent the positive space-time curvature of a hyperspherical universe then it tells us the exact distance to the antipode (effectively the 'size' of the universe) and also its exact weight, because a simple equation links together these quantities for a structure with an orbital velocity of lightspeed.

The measured value of the Anderson deceleration gives an antipode distance of 11 billion light years, and this represents the greatest separation that any two points can have in a hyperspherical universe

This cosmic deceleration factor arising from the spacetime curvature offers an alternative explanation for redshift, which in a hypersphere results simply from distance, not from a general expansion of the universe.

The geometry of a hypersphere has an additional lensing effect which tends to magnify objects in the vicinity of the halfway to antipode distance and to reduce the apparent size of objects further away. This explains why the redshifts of the type 1A supernovae used as 'standard candles' do not match distance estimates derived from apparent magnitude. This mismatch has led to the erroneous conclusion of an accelerating expansion of the universe, and the hypothesis of dark energy to propel it.

The vorticitation of the hypersphere implies an omnidirectional type of rotation in which all widely separated bodies rotate around each other, and this rotational frame adds significantly to orbital velocity at galactic distances by a factor of the square root of distance times the Anderson deceleration. At planetary distances the effect remains negligible, but at galactic distances it allows orbital velocities to have higher than expected values, without dark matter.

A hypersphere has a finite and unbounded extent in space. You cannot get out of it because it has an orbital velocity of lightspeed, and an unachievable escape velocity of the square root of twice lightspeed, yet you can travel around in it as far as you like without encountering any kind of edge or boundary. If the universe consists of a hypersphere then the question of what lies outside of it has no meaning because all of the 3 dimensional space that exists lies within it. Space does not consist of the mere absence of stuff, it consists of the curvature subtended by matter, and where the matter ends, not even space exists, so it has no outside. However a hyperspherical universe will have a spatial horizon, a distance beyond which you cannot see anything, because light from objects near your antipode will become redshifted to oblivion, and the antipode will appear to lie at the extreme of every direction you look in, rather as the south pole of the earth lies in every possible southward direction from the north pole of the earth.

The hypothesis of 3 dimensional time advanced in chapter 5 of course applies to the universe as a whole and the positive spacetime curvature arising from the gravity of the universe would also lead to a universe with finite and unbounded extent in time. Thus although the universe will exhibit a temporal horizon of 11 billion years, nothing in principle prevents something from persisting for longer. Some of the older galactic structures do seem to have an age greater than the temporal horizon, but few of the macroscopic structures in the universe seem likely to survive for such lengths of time.

Stars expand and explode consuming entire planets and heavy neutron stars recycle higher elements back into hydrogen.

Thus the helium abundance does not require a primordial fireball to explain it. The proportion of elements in the universe represents a constant equilibrium.

Light from a distant galaxy that comes towards an observer will become redshifted by the Anderson acceleration. However, light from objects which travels away from the observer will also eventually reach the observer, having passed the antipode and come back again. This light will have travelled more than once round the universe and become profoundly redshifted. Yet it will not completely disappear because the vast tracts of space it passes through contain diffuse clouds of gas and dust which gradually absorb and re-emit the light until it reaches equilibrium with the temperature of the dust and gas in intergalactic space. Absorption and re-emission begins to dominate over the effect of further redshifting as the lights energy drops towards that of the temperature of the intergalactic medium. This residual light then appears to us as the microwave background. It does not represent the cooled afterglow of a cosmic fireball, it merely represents the constant temperature of the universe, which comes in at a rather chilly 2.7 degrees above absolute zero, because it mostly consists of rather cold and fairly empty space.

Part 3 Hyperspherical Metaphysics

Although the hyperspherical universe outlined above has a spatial and temporal horizon beyond which we cannot see; it has no beginning or end. Although both space and time exist as vast closed curved structures, events within this universe do not undergo eternal recurrence. If you sit still for 22 billion years you will in theory return again to the same point in spacetime in this vorticitating structure, but don't expect to find the exact same events occurring there again, because events will have moved on.

The hyperspherical universe hypothesis gives rise to a peculiar inversion of the type of answerable question that we can pose. We can ask and answer the question of why it exists. It exists because it has self-consistency. However we cannot ask or answer the question of how it got that way. We have a strange tendency to regard nothing as somehow more fundamental than something. Yet we have absolutely no reason for this assumption, indeed the evidence all points to the contrary. We never observe anything coming from nothing, everything we observe appears to have come from something else. Structures come and go, but the underlying space, time, mass, and energy merely rearranges itself endlessly. We can observe no mechanism which creates these phenomena, nor any which could lead to their demise, so why do some people persist in imagining that the universe has an origin from some prior state of nothing? I suspect that the whole idea arises from our lamentable capacity to ascribe reality to things that don't exist like 'being' and to privative concepts like 'nothing'.

So does the hypothesis of a vorticitating hypersphere constitute a TOE, or 'theory of everything'? Most definitely not, and it seems that Gödel's Incompleteness Theorem prevents any sort of TOE from existing, because it proves that any system of maths or reasoning must contain assumptions that we cannot derive from the system itself.

It does however provide a more elegant model of the cosmos than that given by the standard big bang. It depends on only four parameters, G, the gravitational constant, c, lightspeed, h, Planck's constant, and A, the curvature parameter. The relationships between these constants define the sizes of particles and associated forces, and the size of the universe itself. (The electroweak and nuclear forces seem to arise from rather complex relationships in rotating 6 dimensional quantised spacetime which require further explication).

However we cannot derive G, c, h, or A from the hypothesis itself, or from each other, we have to measure them. The hypothesis remains incomplete because we cannot tell why these constants have their observed values, although the word 'observed' may in itself provide a clue.

Nothing 'is' true, but the most self-consistent hypotheses have the greatest utility until someone uncovers incompleteness or finds a more fundamental assumption. Chapter 7 explores the possibility that Psi, the psychism in panpsychism, may supply the missing ingredient.

Part 4. The Map, the Journey and the Meaning

'The dimensionality of the map one uses depends upon the journey being undertaken' - Waldo Thompson.[18]

Flat Earth theory serves well enough for a trip from the cave to the water hole and back, and a third dimension going up into the sky and down underground serves to accommodate gods and devils.

A lot of people still think like that, believe it or not.

Spherical Earth theory serves well enough for trips to other continents and gives some intimation of the great space beyond. The gods and demons begin to retreat into unseen dimensions.

Flat Space theory serves well enough for trips around the solar system if you acknowledge gravity as a force. Those pictures

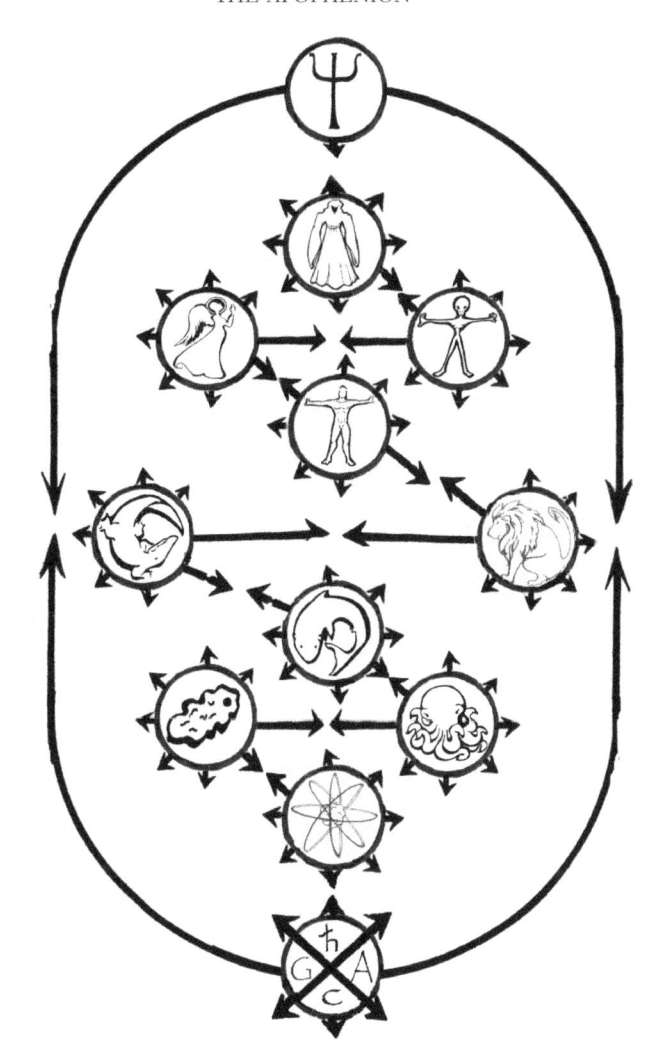

A Chao-Panpsychic Tree of Life.

Herewith some arbitrarily selected steps on the way.

From the perspective of level 7 look down for shamanism and science, look up for religion and mysticism, and for magic look in all directions, nobody knows where most of the arrows go.

Level 0. Some of the fundamentals.

Level 1. Atoms, matter self-organises.

Level 2. Unicellular life, an Amoeba, our GreatGrandparent.

Level 3. Invertebrates. Lord Cthulhu presides. Life gets nasty.

Level 4. Vertebrates. Still nasty, but quicker and a bit smarter.

Level 5. Reptiles. Out of the water, but still in our hindbrain.

Level 6. Mammals & Birds. Neat tricks, they can go everywhere.

Level 7. Us. Still half ape and part crocodile, but dreaming of improvements.

Level 8. Angels and gods. Our dreams of improvement, mostly foolish.

Level 9. Aliens, barely imaginable advanced versions of ourselves.

Level 10. Unimaginably advanced forms of life.

Level 11. Psi. Cosmic Panpsychism.

The Kabbalist may prefer to view the tree as top down; the Scientist may prefer to view it as bottom up, the Panpsychist reserves judgement.

of the Earth from the Moon were worth a thousand words about what it means to live on a planet in a space of almost indescribable enormity. The gods and demons have no place to hide but in the hearts of humans.

Curved Spacetime theory leads to an apocalyptic universe with a beginning and an end, ruled either by blind chance or an absentee landlord who lives elsewhere. The geometry of this map effectively prevents us from ever travelling far in the territory.

Vorticitating Hyperspherical Spacetime has no beginning or ending but its finite and unbounded extent does not render it incomprehensibly infinite in space either. The chance which rules it does not act completely blindly because 'mind' forms an integral part of its function. Welcome to the participatory universe, the geometry of this map permits magic and invites us to become apprentice gods.

I also suspect that this map will also somehow allow us to take trips right round the territory eventually.

Chapter 7 Illumination?

Throughout recorded human history some people have always sought some kind of transcendence in the idea of gods, or higher states of 'being' or in expectation of after death states in which they somehow achieve union with something far greater than themselves.

Mostly this has led to ghastly disaster here on earth.

Nevertheless such ideas stand as a tribute to the power of imagination and an insult to the theories of cybernetics. (At least one species of organism in this universe can imagine a greater state of complexity than it posses itself, even if it usually comes down to fantasies about bigger penises or greater destructive capabilities, or merely some elaborate excuses for burning a few enemies at the stake.)

This chapter seeks an Apophenia in the most despised of all the classical arguments for the existence of the gods, The Ontological Argument, which basically says that if we can imagine them, then they probably exist.

Part 1. A Fifth Principle of Thermodynamics?

Note that a Zeroth law of thermodynamics got officially added for the sake of technical completeness, as the first one didn't seem quite fundamental enough on later reflection, so we can call any new one the fourth or fifth law according to taste.

The philosophically significant second law says that everything runs down towards increasing entropy. Energy dissipates, stuff just falls to pieces with time, it all ends up as an inactive soup of particles at the same temperature with nothing much happening.

Life on earth for example does not really depend on energy from the sun. It depends critically on the sun having a much higher temperature than the surrounding space. Life exists here because it exploits the energy difference between the sun and space. It absorbs the relatively high grade solar energy and excretes the lower grade heat back out to space in a more entropic form. If we had a uniformly warm sky instead of a generally cool sky with an intensely hot spot in it, then life could not exist.

Life here has developed ever more complex and exotic mechanisms for dissipating energy. Herbivores dissipate energy far more quickly than the plants they feed on, carnivores dissipate the energy of herbivores far faster than the herbivores do themselves. Humans dissipate energy at an astonishing rate. Not content with merely eating the plants and the herbivores and the carnivores they also dig up the remains of old plants and animals in the form of coal and oil and burn those as well. Recently they discovered that they could even burn the uranium bearing rocks forged in the death throes of the previous star in this part of the galaxy.

Life dissipates energy and develops ever more complex ways of doing it. It takes a huge area of sunlight absorbing vegetation to maintain a vast number of insects to keep a smallish number of rodents and birds in business, just so that a single family of hawks or eagles can exist.

The second law of thermodynamics perhaps lacks global or cosmic applicability in two important ways. In Biology it fails to account for a tendency towards increasingly efficient and baroque forms of energy dissipation. The definition of entropy remains far from robust, and the relationship between entropy and the amount of information or sophistication in a system remains questionable.

Stephen Hawking brilliantly observed that entropy increases with time because we measure time in the direction in which entropy increases.[15] Thus the second law of thermodynamics constitutes a tautology.

Some theorists have tentatively proposed, as a sort of extra law of thermodynamics, that 'Energy dissipating structures will naturally tend towards more efficiency and complexity wherever possible', mainly on the grounds that they already appear to have done so in evolution here on earth.

On the cosmic scale, entropy may not necessarily constitute the inevitable fate of the universe. The second law of thermodynamics works well enough for steam engines where chemical and kinetic phenomena dominate, but on the larger scale other forces prevail. Gravitation and nuclear forces may well recycle the thermonuclear ash of the heavier elements back into primeval hydrogen when stars collapse. Black Holes and spacetime singularities represent a sort of entropy rich dead end in the evolution of the universe, but I suspect that either neutron annihilation or the constraints of lightspeed orbital velocity prevent them from forming in reality.

Part 2. What Can Have Evolved?

Although the universe may have an 11 billion-year temporal horizon, you can go around the temporal curvature as many times as you like, if you have the technology and the will to survive. Life thus effectively has, and has had, unlimited time at its disposal.

If some kind of extra law of thermodynamics does favour the evolution of increasing sophistication and complexity of life in the universe, then it follows that the most sophisticated intelligences that this universe can possibly support must already exist, and probably in very large numbers.

Part 3. Science Fiction Gods

Do they take much of an interest in us? I doubt it. How much entertainment does an ant's nest provide you with?

'Adepticus Sir, that bunch of Ornithoids on Arctos 4 that you asked me to observe, well they've just trashed their planet'.

'Oh that is a pity Initiatus Jones. What was it this time, ecological screw up or nuclear winter?

'Worse than that Sir, it looks like they were mucking around with vacuum energy without having first invented the Mobius sphere'

'Ah yes, the old classic mistake, we loose a few like that'

'Could we not have tipped them off about it Sir?'

'I'm afraid not Jones, stupidity must remain its own reward, it's regrettable but there you are, did you salvage anything?'

'They composed some fairly good poetry a couple of centuries ago, and some rather fine cloud sculptures fairly recently, I've logged some records in the archives'.

'Splendid Jones, I'll peruse them this evening. What about those Apes on Sol 3, how are they getting on?

'Quite a bit of warfare as usual Sir, mostly based on chemical explosives these days, but with the occasional use of plutonium. Many of them have developed a belief in a big bang theory, and they reckon that they have the maths to prove it'

'Really? Smith in anthropology will probably find that hilarious, I'm sure she would appreciate the data. It was one of her old stomping grounds you know'.

'No I didn't know that Sir'

'It was a long time ago Jones, and bit of a fiasco actually, she gave them a piece of her mind about some of their barbaric behaviour which then abruptly became worse. Ever since then they have been obsessed with the number plate on her craft, it read JHVH. The department gave her a desk job after that.'

Many 'ifs' surround the whole question of intelligent life in the universe but only one of those 'ifs' really counts.

If the physics of this universe absolutely prevents communication or travel between star systems, then it does not matter how much intelligent life exists, it can never affect us, and we shall eventually become extinct when our star starts getting low on fuel.

On the other hand if intelligent life can break free of the star systems in which it develops, then the universe must swarm with intelligent organisms. Life went into a bit of a funk here on earth for hundreds of millions of years as massive reptiles plodded about doing nothing very interesting for a very long time. Intelligence only has a history of a half a million years or so here. On other worlds dumb slugs may still gnaw the vegetation billions of years down the line, but if intelligence develops on only a minuscule fraction of worlds, then the universe must still contain a vast and varied resource of it. Statistically, a fair amount of it must have far greater abilities than we have yet.

Do highly evolved life forms take much interest in less advanced life forms such as us? Well we cannot know their motivations, but curiosity seems an indispensable attribute of intelligence, so would we seem interesting enough to warrant their attention? I very much doubt that any of our science and technology would interest them in the slightest. If they have the capacity to come here, or to examine us remotely, then all of our technology would seem laughably primitive to them.

Perhaps some of them might have an interest in our cultural activities for academic or entertainment purposes. Maybe some like watching primitive battles or our attempts at art or magic, perhaps their anthropologists find our attempts at religion an hilarious reminder of their own culture's long distant foibles and delusions.

Do they ever intervene in the development or survival of less advanced species? I would suspect that in general they avoid doing so. If we interest them in any way at all, we would become less interesting the more they interfered.

Nevertheless it remains possible that highly developed intelligences of extra-terrestrial origin do sometimes take an interest in the activities of humans. Maybe on very rare occasions they do intervene, but perhaps only with the same sort of random whimsy that you or I might move a snail with a particularly attractive shell off the pavement onto someone's front lawn.

It seems highly probable that highly advanced intelligences have already evolved in the universe. It seems unlikely that they will offer us much help here on this little ball of rock, and more likely that they want to see what we can make of ourselves by our own efforts.

Let us not disappoint them, or ourselves.

Part 4. A Panpsychic Universe?

At the time of writing, quantum-cosmology looks like a grotesque mess.[19, 20]

We cannot specify why the observed physical laws and constants take the form and the values that they do. We understand many of the laws of the universe but we have no idea why they exist.

If the various constants like the relative masses and charges of fundamental particles had even fractionally different values then life would not exist in the universe. Stars would either not form or they would burn too quickly and the rich chemistry which supports life would not happen with any other conceivable combination of values.

We seem to inhabit a 'Goldilocks Universe', not too hot and not too cold, and replete with the perfect chemical porridge to support life.

This has led some theorists to assert an Anthropic Principle which basically states that the universe looks precisely like this because if it didn't, we wouldn't exist to remark upon the fact. That at least seems unarguable.

Yet the inability of conventional physics to specify any reasons for the existence of this particular set of prevailing laws and constants has led to some highly dubious speculation about a meta-universe or 'Multiverse' of which this observed one forms only a microscopic fragment.

In some Multiverse hypotheses new universes can somehow become created from black holes within existing universes. Black holes supposedly collapse into singularities which erupt 'somewhere else' as big bangs which then initiate new universes with randomly selected new laws and constants Thus the number of universes tends to multiply hugely with time and perhaps some kind of Darwinian survival of the fittest universes applies, as some of them may collapse quickly or fail to form black holes to birth new universes. Alternatively, in simpler versions this universe periodically collapses in a big crunch and out of the resulting singularity a new universe explodes into existence in fresh big bang with a new suite of laws and constants. We just happen to live in one of the incredibly rare editions that can support life.

Such hypotheses have developed partly because String and Brane theories, which attempt to account for fundamental particles in terms of a spacetime geometry which has many extra small spatial dimensions, all yield fantastic numbers of possible answers, very few of which correspond to our observed reality.

Both versions of the Multiverse theory seem to severely violate the principle of Occam's razor in their attempt to merely account for the inability of theorists to specify reasons for the laws and constants of the universe we observe.

Singularities remain unproven, and if universes continually bud off daughter universes where does the mass and energy for their formation come from?

What meaning can the 'somewhere else' that these new universes supposedly manifest into possibly have? What keeps them gravitationally isolated from their mother universes?

A simpler solution may lie in applying General Metadynamics to the Fifth Principle of Thermodynamics and then adding Panpsychism.

Life then ensures the conditions for its own development in a single universe.

In this model only one universe actually exists, and it inevitably contains life because circular time and retroactive causality allows life to select the conditions in which it can exist.

Some guises of the Muse,
The Chaomeras
Pareidolia, Apophenia, and Eris.
Three Wyrd Sisters of Chaos,
Pareidolia making augury from entrails,
Apophenia seeking mysterious connections,
Eris disordering our carefully crafted expectations.

Thus in a very real sense we would all comprise the 'God' that specifies the universe.

Atoms and molecules and phenomena with a simple structure presumably make a small contribution to it, perhaps we make a larger contribution but we should not delude ourselves with ideas of omnipotence here, because the universe probably contains more sentient races than individual humans.

A Panpsychic Universe would represent a collective effort by the entire Mind behaviour within it.

$G, c, \hbar, A, \Psi$.

Water, Air, Earth, Fire, 'Spirit'?

Well at least that conforms to Eris' Iron Law of Fives.[21]

This perhaps explains the astonishing diversity of the universe's contents and phenomena, including the unpleasant bits.

Chapter 8 An Invocation of Apophenia

Part 1. Introduction

Apophenia means finding meanings and connections where others have not; it thus underlies both psychosis and genius. Its occurrence has created progress and innovation in many forms of human mental endeavour. Apophenia has a sister, Pareidolia who brings visions where others see nothing. Whereas Apophenic insights tend to help in magic and science, Pareidolic insights tend to fuel art and religion.

In most disciplines, Apophenic advances arise fortuitously and accidentally, and the disciplines themselves contain no formal procedures for inducing it, practitioners just hope that imagination and intuition may eventually kick in. Art however has recently experimented with various stochastic techniques, the random fall of paint or the random literary cut-up provide recent examples.

The majority of Apophenia inducing techniques actually come from magic and the occult because of their association with sortilege and divination and forbidden realms of enquiry.

> Kabbala began as a technique for inducing Apophenia.
> (She told me that Herself)

The ancient Hebraic sages attempted to find extra meanings and inspirations in their scriptures by assigning numerical values to letters, words and phrases and then looking for arithmetically equivalent words and phrases. Of course with the passage of time the resulting insights became ossified as 'divine maps' of various kinds; and creative use of Kabbala tended to dry up, although interesting revivals of the technique have appeared in various eras. The world owes a considerable intellectual debt to the genius of Hebraic thought in many fields.

The Moslems also had a kind of Kabbala based on the Zairja, a series of rotating discs inscribed with the letters of the Arabic alphabet which they turned to create new combinations of ideas and concepts.

Writing in 13th Century Spain, Ramon Lull developed his Ars Magna, a technique for randomly combining concepts using stacks of progressively smaller rotating discs with words and symbols on them. For this he almost certainly took some inspiration from the Zairja that he would have encountered on various missions to North Africa.

Ramon Lull's Ars Magna devices carried mainly theological and philosophical ideas and symbols, and as with any computer, if you put garbage in, you get garbage out. Nevertheless the technique itself created enduring interest, and centuries later that giant polymath of the early scientific age, Gottfried Wilhelm Leibniz, used it as the basis of his De Arte Combinatoria.

Ramon Lull also wrote the original Liber Chaos. Reading between the lines of this strange tome one cannot but conclude that he regarded Chaos as more fundamental than any God, rather as the ancient Greeks did. However, Lull lived under the shadow of the Inquisition and he came under suspicion at various times. Under such circumstances one had to write with a certain circumspection and circumlocution, or face the stake. Amazingly, Lull managed to remain more or less in the favour of the church powers, and they even preserved his deeply heretical Liber Chaos for him, not having the imagination to understand what he was implying. He acquired the informal title or nickname of Doctor Illuminatus.

The graphic representation of concepts and ideas and their geometric relationships has become a staple tool of thought, but the random combination of such concepts and ideas remains rather esoteric, yet Dynamic Ideational Geometry, as we can call it, provides a tremendously powerful and useful tool for inducing Apophenia in more or less any discipline. It forms the basis for the following approach to Invoking Apophenia.

Part 2. General Observations

The operator can invoke Apophenia on any subject and with any desired degree of intensity. A mild invocation may prove useful for solving particular problems with eccentric insights and need consist of no more than some work at a desk followed perhaps by a walk in the woods. A more intense invocation might consist of an elaborate ritual, cut -up incantations, disinhibitory or hallucinatory sacraments, and intense meditation on strange glyphs and diagrams, and deliberately induced sleep pattern disturbances. This may well leave the operator mentally hyperactive and somewhat disturbed, and possibly somewhat pareidolic, so a formal banishing can follow an intense working. The banishing itself may well work better if followed by deliberate re-immersion in mundane activity, particularly physical work.

In more intense workings, magicians may wish to conceptualise themselves as Apophenia in person rather than simply as an abstract principle.

Plato got it wrong when he identified Necessity as the Mother of Invention.

Very rarely can we invent anything to order. Most inventions come when someone finds an inspiring connection between existing ideas and gives birth to another, so we must regard Apophenia as the real Mother of Invention.

In terms of Chaos magic symbolism, Apophenia has a Uranian quality. Uranus lies outside of the orbit of the seven classical heavenly bodies that represent ordinary drives and motivations. It provides a counterpoint to the central Solar ego or normal personality. We find Apophenia out in the darkness beyond known knowledge, at the frontier between what we know and what we can perhaps intuit or imagine.

She represents an alternate mode of entry to Uranian magic that complements the rather more macho god form of Ouranos who seeks to force the gates of the beyond with strange antinomian conjurations and tries to impose form on what he finds there.

Apophenia just opens the gates, and delights in what comes out. Sometimes on the other side of the gates her crazy sister Pareidolia awaits her, at other times Eris the goddess of discord appears to throw paradox and confusion into the works, just to stir things up. Beware the three Weird Sisters of Chaos, they make challenging Muses.

The symbol of Apophenia shown in her hand consists of five elements, a cross, a circle, and three crescent moons. These combine to include the currents of Uranus, Sol, Luna, and Venus, with a suggestion of Mercury.

Part 3. The General Form of the Invocation

Magicians will need to spend some considerable time and effort in the preparation of the materials and concepts needed to support the birth of a goddess within their psyches. She has only existed as a god form since 2005 and she needs all the support her Priests and Priestesses can give her, but she gives much back in return.

The old Grimoires demanded considerable efforts at exacting preparations for good reason. It takes time and thought for imagination and belief to build up to useful levels.

An Apophenia Wand and tables of Dualities, Trialities, Quadrads, Pentads, (and higher order figures, if desired,) need preparing in advance. Some examples of tables appear below. Magicians should also construct an Astronomicon, and they may well supplement the period of instrument preparation with practice in consuming the sacrament mentioned below, to acclimatise themselves to the taste and effects. The courteous Magician should also acknowledge the possible presence of Apophenia's sisters in the ritual and prepare a symbol of Eris (see figure) and the materials to create a Rorschach Blot to welcome Pareidolia, and place them at the extremities of the altar.

Perchance the magician may need to add something to the tables or record an insight, writing instruments may also adorn the altar desk.

Magicians usually perform this invocation alone although work with the tables on a suitable altar or desk can take the form of a quickfire brainstorming word association exchange between two or more operators.

Fashion the tables from stiff paper or card. Fashion the wand from any material, a little longer than a hands length. A wand cut from a thick sheet of Aluminium serves particularly well, the symbolism of this light, amphoteric, versatile, and reactive metal proves particularly germane and the result should easily repay the efforts with hacksaw and files.

An Astronomicon typically consists of a black disc of at least a hands length in diameter and upon it the magician moves smaller appropriately coloured discs to represent various archetypes symbolised in planetary form. A Steel disc, enamel painted in matt black, serves well as the void of space. Magnetic discs painted to represent the seven classical planets plus Uranus then serve well as the minimum number of movable pieces.

The full Apophenia invocation begins with a banishing ritual if required (the Gnostic Pentagram Ritual serves well here).

A statement of intent (to taste) begins the ritual proper.

The Magician then delivers a spontaneous appeal to Apophenia delivered verbally or mentally in the vernacular.

Taking the wand, the Magician draws the symbol of Apophenia in the air or smoke and then visualises drawing it in to suffuse the entire physical body. Rapid breathing to hyperoxygenate the brain often proves useful. Magicians may employ supplementary forms of Gnosis such as erotic or autoerotic excitation at will.

The magician then delivers a 23 word invocation in Uranian-Barbaric, previously committed to memory.

The Magician (as Apophenia) then welcomes Eris whilst gazing at her symbol and contemplating briefly the clash of opposites.

The Magician then welcomes Pareidolia by making a Rorschach Blot and contemplating the result.

The Magician then politely requests that these goddesses to remain on the periphery of the ritual

Incense, if required, should consist of a mixture of agreeable and disagreeable ingredients. An Oakmoss and Valerian root mixture serves particularly well.

The alkaloid Theobromine (Xantheose) forms the basis of any sacrament to Apophenia. Prepare a very strong decoction of Theobroma Cacao (Cocoa) in hot water. The goddess loves the chocolate alkaloid, but chocolate confectionary consists mostly of fat and sugar with precious little active ingredient.

The Magician then begins work with the prepared tables, pointing with the wand at various of the figures in the tables as they catch the attention. The horns of the moons on the wand can serve to form a symbolic bridge between concepts. The magician can repeat the Ouranian-Barbaric Invocation at will, or use it as a continuous chant.

At various random or inspired intervals during the work with the prepared tables the magician may turn to the Astronomicon and manipulate the moving pieces to create an additional stream of consciousness or as distraction from one which has become blocked. Contemplate the flavour of such conjunctions as solar-martial thought, or lunar-jupiterian attitudes, or mercurial-saturnine philosophies, or whatever may arise by chance or design.

Work continues till exhaustion or inspiration supervenes. Inspiration may come to fill the vacancy attendant on relaxation after exhaustion, so use a final banishing only if disagreeable phenomena persist.

In summary, the full Invocation proceeds as follows, with improvisation and amendment on inspiration:

i) Banishing ritual if desired.

ii) Ignite incense and consume sacrament (if desired).

1) Statement of Intent.

2) Appeal to Apophenia.

3) Draw and visualise and suffuse oneself with the Apophenia symbol.

4) Apophenia incantation in Ouranian-Barbaric

5) A nod to Eris and Pareidolia.

6) Work with Tables and The Astronomicon.

7) Banishing ritual if necessary.

The Apophenia Invocation

The Invocation appears as 23 words of the Ouranian-Barbaric magical language with an approximate English translation italicised below. Those who desire to maximise its efficacy in use should commit the Ouranian-Barbaric phrases to memory by repeated chanting, until it flows fluently, but they should avoid consciously learning the vernacular (English) meaning of it.

Having read the vernacular form several times, the magician should obliterate it from the page.

XIQUAL UNGASCAB GESIZAL CHUWAKAGATHAZ CUDTEG

Phenomenising Uranus Goddess Chaos Lady

COYANIOC FODDAWITH POZATHOR GYCAPORUS GODON

Join together Stokastic Reality, Random Illusion

CHAEQUAI NEKOZY CHAZITER EMUUL ETHENG

Entangling Imagination Coincidence, Do Sex, Do Death.

QYOPAL JOACHABIM DOHBLE THECJECHED DAHZOO

Illuminating Intuition, give me Neither-Neither Genius

KABOTHEYA OFTALA AEPALAZAGE

Bring about the latest Octarine End of the World

A Table of Dualities

The magician may add or subtract from the following list at will and inspiration.

The following merely provide some examples of useful starting points drawn largely from magic, mysticism and physics.

The magician concentrates upon chosen dualities by The Neither-Neither technique.

First consider one side of a duality on its own, and then the other, then upon a conjoining of the two, and then upon the simultaneous absence of both, to see what arises therefrom.

> Doing - Being
>
> Will - Perception
>
> Causality - Randomness
>
> Sex - Death
>
> Fear - Desire
>
> Love - Hate
>
> Ego - Enlightenment
>
> Baphomet - Choronzon
>
> Eristic delusion - Aneristic delusion
>
> Atman - Annata
>
> Space - Time
>
> Mass - Energy
>
> Science - Magic
>
> Religion - Art

A Table of Trialities

The Magician concentrates on the concept at each of the vertices of a chosen triangle in turn, and then considers how they may give rise to each other in clockwise or anticlockwise sequence.

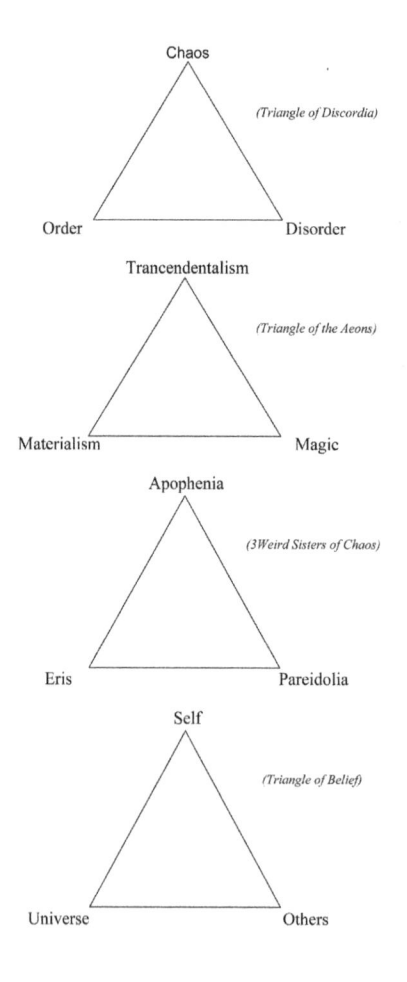

```
            Chaos
            /\
           /  \          (Triangle of Discordia)
          /    \
      Order----Disorder

        Trancendentalism
            /\
           /  \          (Triangle of the Aeons)
          /    \
    Materialism----Magic

          Apophenia
            /\
           /  \          (3 Weird Sisters of Chaos)
          /    \
       Eris----Pareidolia

            Self
            /\
           /  \          (Triangle of Belief)
          /    \
      Universe----Others
```

A Table of Quadrads

The Magician creates Quadrads by crossing pairs of Dualities, and then concentrates upon them as though they represented graphs with the dualities as axes. The magician aims to try and find meanings for each of the four quadrants.

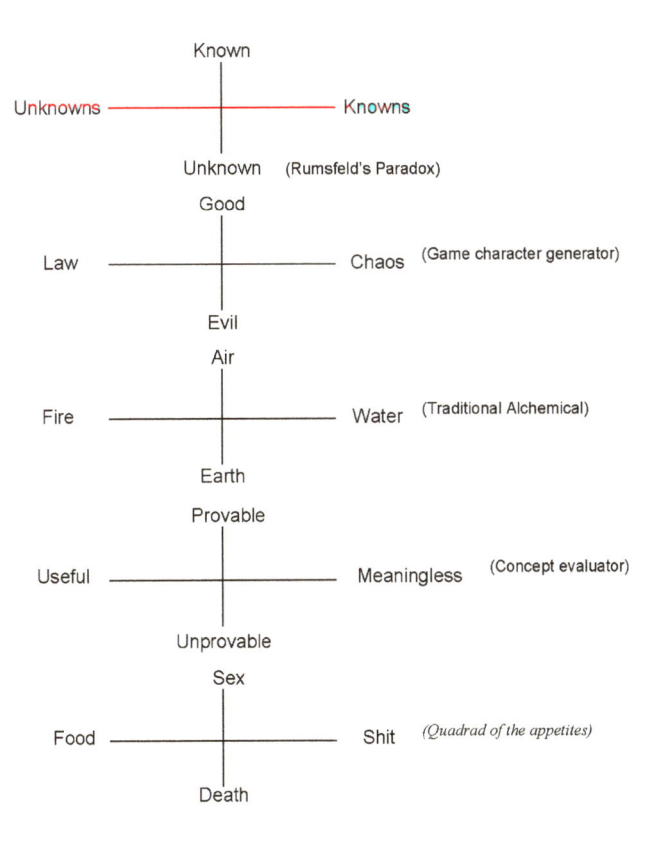

```
                     Known
                       |
      Unknowns ————————+———————— Knowns
                       |
                    Unknown    (Rumsfeld's Paradox)

                     Good
                       |
          Law ————————+———————— Chaos    (Game character generator)
                       |
                     Evil

                      Air
                       |
         Fire ————————+———————— Water    (Traditional Alchemical)
                       |
                    Earth

                   Provable
                       |
       Useful ————————+———————— Meaningless    (Concept evaluator)
                       |
                   Unprovable

                      Sex
                       |
         Food ————————+———————— Shit    (Quadrad of the appetites)
                       |
                    Death
```

A Table of Pentads

All things obey The Law of Fives, and you can obtain any number by mucking about with 5, for example, 5 = 3 + 2, and then 3 - 2 = 1, and from then on to any number desired. Moreover, five represents that sort of divine spark or awkward extra bit that lies in excess of foursquare ordinariness.

Five therefore appeals to Magicians and antinomian-minded people everywhere, probably more than any other prime number.

So wherever you see 4, look for something to complete The Iron Law of Fives.

Some examples follow:

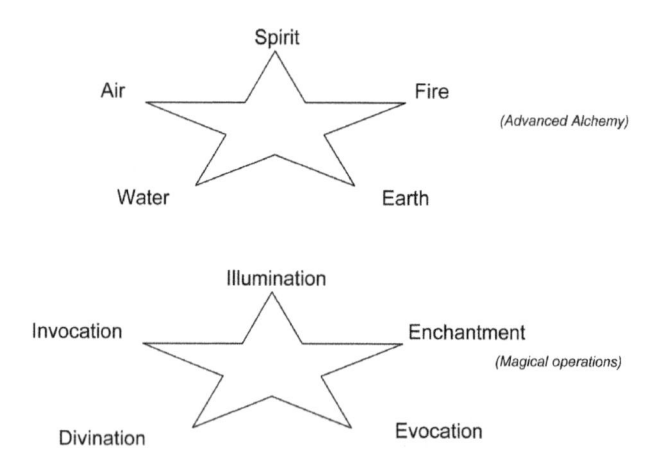

```
           Spirit
    Air  ───△───  Fire
         ╲ ╱ ╲ ╱              (Advanced Alchemy)
          ╳
         ╱   ╲
    Water     Earth
```

```
         Illumination
 Invocation ───△─── Enchantment
            ╲ ╱ ╲ ╱           (Magical operations)
             ╳
            ╱   ╲
      Divination  Evocation
```

The magician may often discover fresh pentads by meditating upon what may lie on an axis going through the plane of a quadrad.

Appendix I Three-dimensional time and quantum geometry

Part 1. The Prologue to a Quantum Geometry

Two theories describe the four fundamental material forces that seem to characterise this universe at the time of writing.

The theory of General Relativity describes how gravity works in terms of spacetime curvature, and this seems to work fairly well, and rather more precisely than Newton's theory of gravity, when it comes to working out how things interact with big objects like planets and stars. However it doesn't seem to give correct answers for the behaviour of whole galaxies and its predictions for the whole universe remain rather open ended.

The Quantum theories describe how the strong nuclear force works (this holds the nuclei of atoms together), and the electromagnetic force (this controls how atoms behave chemically and how they interact with light). They also describe the weak nuclear force which theorists now regard as specialised aspect of the electromagnetic force, so they tend to refer to a single electro-weak force nowadays. Quantum theories model these forces as mediated by 'real' particles and fields that supposedly consist of 'virtual' particles.

Unfortunately the Relativity and Quantum theories do not fit comfortably together, indeed they seem to contradict each other completely in principle. Relativity implies a continuously divisible and ultimately causal and determinate universe with strict temporal and spatial locality which does not allow anything to exceed lightspeed or to travel backwards in time. Thus relativity remains an essentially classical theory in which we can model the universe geometrically, even though we have to accept that a large concentration of mass or energy, or an extreme acceleration can distort the geometry of spacetime. (The earlier and simpler theory of Special Relativity describes how velocity alone can create spacetime distortion).

Quantum theories on the other hand imply that nature does not exhibit continuous divisibility, at some point we must encounter the smallest possible pieces of mass and energy and probably of space and time as well. Moreover the quantum perspective implies that the usual classical rules of causality and locality do not apply, or at least not very strictly.

For over seventy years theorists have attempted to reconcile the underlying conflict between these two rival descriptions. The conflict goes beyond physics into the realms of metaphysics, the realm of our basic beliefs about how reality actually works in principle. Because quantum theories can model three out of the four fundamental forces, attention has tended to focus on developing a quantum type theory of gravity. This quest has so far proved fruitless, the supersymmetry particles predicted by the simplest quantum gravity theories have failed to appear in experiments. The more sophisticated Superstring and Brane theories have failed to produce testable predictions, and the quantum gravity particles theoretically responsible for mass and gravity, the Higg's Boson and the Graviton, remain undetected.

Thus perhaps we should consider geometricating the quanta instead of trying to quantise gravity.

Three dimensions of time, plus curvature, together with the accepted three of space, plus curvature, seem to provide exactly the required degrees of freedom to accommodate the known suite of particle behaviours. In this model, particle properties arise from rotations of the three spatial and the three temporal dimensions about the fourth (curvature) axes of space and time.

Part 2. Fundamental particles in eight dimensions

In this model called Hyperspin Eight Dimensional, or HD8 for short, the six space and time axes of a fundamental particle can rotate through the fourth dimensions. As all eight dimensions lie orthogonal, (at right angles) to each other, the spatial and temporal axes can rotate relative to either the spatial or temporal fourth dimensions

I do not know 'what' actually spins, but I suspect that fundamental particles consist of the quanta of spacetime itself somehow endowed with spin. This quantisation appears to occur at the level of the so-called Planck scale, of about 10^{-33} metres and 10^{-44} seconds, so fundamental quanta appear as virtually zero size points in particle mode.

We can designate the dimensions of space and time as $s1$, $s2$, $s3$, and $t1$, $t2$, $t3$, and the fourth curvature dimensions as $s4$ and $t4$.

Anchoring the rotations on the curvature axes explains in principle the origin of mass and gravity, for spacetime curvature corresponds to what we perceive as mass and gravity. Increasing the number of axes rotating about the fourth dimensions generally increases the mass of the fundamental particle as the rotations act as a store of energy, however no simple algorithm for particle masses arises from this model as yet.

Complete rotations relative to the fourth (curvature) dimensions of space and time have the effect of making a 3D object turn into its mirror image and back again.

Consider a six-sided dice. Swapping over the faces marked six and one creates a mirror image of the original dice which no kind of rotation in three dimensions can restore to its original form. Similarly, swapping all three pairs of opposite faces also creates the mirror image of the original dice.

Swapping any two pairs of opposite faces however merely has the same effect as rotating the dice in three dimensions. We can see this effect manifest in the suite of observed fundamental particles; none of them exhibits two axes of the same type rotating against one of the fourth dimensional axes on their own.

The dice analogy does fail to show a particular feature of rotation in a fourth dimension, it can occur either clockwise or anticlockwise in the fourth dimension, even though the result looks the same because the fourth dimension remains invisible to us. Thus the rotations of the six dimensions about the fourth dimensions can each occur clockwise or anticlockwise, corresponding to the positive and negative generational, electroweak, and colour charges.

Consequently the following classes of spin become possible:

4-Axis	Spin	Particle Property
s4	s1 or s2 or s3.	Chiral Spin, + or -
t4	t1	
t4	t2	Colour 'charge', + or - R, G, B
t4	t3	
t4	s1	
t4	s2	Electroweak 'charge', + or - 1, 2, 3.
t4	s3	
s4	t1	
s4	t2	Generational 'charge', + or - 1, 2, 3.
s4	t3	

By applying a few simple rules to the above scheme we can account for the whole suite of observed particles.

1) A particle must have at least one rotation in space and one in time. This amounts to no more than saying that it must create a finite amount of spacetime curvature.

2) A particle must exhibit 't4-axis neutrality' which means that it can only have either zero or +3 or –3 rotations about t4.

3) Bosons (energy particles) consist of particle-antiparticle doublets that have aligned chiral spins, thus giving them twice the spin of Fermions (matter particles).

4) Particles cannot have more than one spatial rotation against s4 or more than one temporal rotation against t4. The s4/s2 and s4/s3 spins denote chiral spins transverse to the direction of propagation. The three spins t4/t1, t4/t2, and t4/t3 denote the colour charges of red, blue and green and their anti-colours when reversed, of which quarks and gluon 'halves' can only carry one.

This simple scheme can model all the particles and antiparticles we observe and also clarify some of their peculiarities. The principle of 't4-axis neutrality' means that electrons have to exhibit 3 units of electro-weak charge, (conventionally denoted as minus 1). The principle applies twice over to quarks. Quarks always have to appear in triplets as hadrons such as the familiar proton and neutron, or as meson doublets to preserve t4-axis neutrality. Quarks also have an electroweak charge of either +or – 1/3, or + or – 2/3 of the electron charge, to maintain t4-axis neutrality as they can only carry one colour charge each. Thus at each generation two types of quark (and antiquark) exist, the familiar Up and Down quarks that make up most of the matter in the universe, and also the Strange and Charm, and lastly the supermassive Bottom and Top varieties.

HD8 does not give a mass algorithm for calculating particle masses but it implies that the addition of spins with increasing charge causes increasing distortion of spacetime and thus requires a higher energy input which appears as mass, although not in any easily quantifiable way.

HD8 does explain the apparent non-conservation of generation in particle interaction. The generational characteristic has spatial reversibility, not temporal reversibility. It also explains the apparent parity violation of neutrinos and the W+ and W- bosons.

All neutrinos appear left-handed and all anti-neutrinos appear right-handed because only the direction of their s1 spins differentiates them. W- bosons consist of electron-antineutrino doublets whilst W+ bosons consist of positron-neutrino doublets.

According to HD8, neutrinos should annihilate in head on collision and liberate energy for new particle creation. The hypothesis also strongly suggests that neutrons behave in the same way at high enough energies, as they have overall colour and electroweak neutrality. Thus Black Holes and singularities do not form in galactic cores, only neutron stars form, and at high densities these stars begin to annihilate neutrons against each other, shedding matter and radiation back into space.

HD8 allows the existence of a wide range of massive and inconsequential bosons that will probably only have a fleeting existence, and it specifically predicts that the Higgs Boson does not exist. Mass arises as an intrinsic quality of particles as a consequence of their fourth dimensional nature.

The suite of known fundamental particles exhausts all possible spin combinations, and mass arises from spacetime curvatures subtended by these spins. The acceleration of charge certainly produces bosons, but I suspect that static fields consist of spacetime curvatures that propagate instantaneously and do not require so called virtual bosons to mediate them.

This proposition seems difficult if not impossible to falsify, even though it apparently contradicts special relativity, yet we could hardly use it for signalling purposes.

Gravitons thus probably exist in the form of a 'neutrino-antineutrino' type bosons caused by cataclysmic mass accelerations such as neutron star collisions but gravitational fields remain the product of spin induced spacetime curvatures, and both strong nuclear and electroweak static fields result from higher dimensional curvatures in spacetime.

Particle Physics buffs may care to adumbrate the spins which characterise each type of particle in the above scheme, the entire chart looks rather large, so I'll just present a few examples:-

Particle type.	Chiral	Colour	Electroweak	Generation
Neutrino	s4/s1	none	none	s4/t1
Electron	s4/ + or − s1	none	t4/s1 t4/s2 t4/s3	s4/+ or −t1
Up Quark	s4/+ or − s1	t4/t1	t4/s1 t4/s2	s4/+ or − t1
Photon (Particle)	s4/+or- s1	none	t4/s1	s4/+ or − t1
(Antiparticle)	s4/+or- s1	none	t4/-s1	s4/- or + t1

(Photon showing both particle and antiparticle components)

Note that the photon consists of particle and antiparticle components, thus it has double the chiral spin of fermions, and no overall electroweak or generational charge.

Part 3. Summary.

The above technical digression hopefully serves to show that the three-dimensional time posited in General Metadynamics also has considerable explanatory power in the field of particle physics as well as in modelling quantum and magical effects.

Strange quarks occasionally feature in reality for the same reason that Magic occasionally features in reality – because reality has 3 dimensional time.

Chapter 6 and its appendix will examine the case for three-dimensional time on the cosmic scale, where it has profound implications for our whole philosophy on such topics as infinity, eternity, creation, eschatology, life, the universe, and the meanings that we may choose to abstract from it.

In passing it seems worth noting that the ratio of any of the six dimensions to its curvature dimension has the value of One to Pi. (See Hypersphere material). Now as an irrational and transcendental number Pi might just supply the chaotic basis for the apparently random collapse of quantum states.

Appendix II
Hypersphere from Radius Excess

Positively curved space has the strange property of having a greater internal radius than an observer would suspect from looking at it from the outside. Thus in a sense a massive object has more space inside it than its outward appearance suggests, rather like those Tardis vehicles of the mythical Time Lords.

To visualise how this can happen, consider a curved space of just two dimensions like the surface of the earth. A small circle drawn on the surface will have a radius r, given by the Euclidian formula

$$r = \frac{C}{2\pi}$$ where C equals the circumference.

However a vast circle drawn on the surface of the earth will have a radius longer than this because it will have to go over the hump created by the curvature of the earth.

A circle around the equator will have a radius of a quarter of the entire circumference.

Now the three dimensional version of curved space does not submit to easy visualisation but a hypersphere or 3-sphere has a similar property an ordinary sphere or 2-sphere. Whereas a 2-sphere has a diameter equal to half of its circumference (in 2-dimensional terms), a 3-sphere also has a diameter equal to half of its circumference (in 3-dimensional terms). This occurs because in 2-dimensional terms we have to measure over the curvature of the earth, and in 3-dimensional terms we have to measure over the curvature of space. This arises because the 2-sphere surface lies embedded in 3-dimensional space, and the 3-sphere lies embedded in 4-dimensional space.

Now Schwarzschild derived a formula from the equations of General Relativity that shows how the mass of any object curves space and leads to a radius excess inside of it. The radius excess depends only on the mass m, of the object and takes the form

$$\text{Radius excess} = \frac{Gm}{3c^2}$$

Where G = the gravitational constant, and where c = lightspeed.

The earth incidentally has a radius excess of only about 1.5 mm, whilst the much more massive sun has a radius excess of about 0.5 km.

The phenomenon of radius excess allows a cheeky little proof that at some state of density, a sphere must become a hypersphere as its radius excess increases its diameter to half of the circumference and beyond.

In the following proof, C = circumference, to which we add radius excess to see at what ratio of mass to diameter, the diameter becomes half of the circumference.

$$\frac{C}{\pi} + \frac{2}{3}\frac{Gm}{c^2} = \frac{C}{2}$$

$$\frac{2}{3}\frac{Gm}{c^2} = \frac{C}{2} - \frac{C}{\pi}$$

$$\frac{2}{3}\frac{Gm}{c^2} = C\left(\frac{\pi-2}{2\pi}\right)$$

$$\frac{m}{C} = \frac{c^2}{G}\frac{3}{2}\left(\frac{\pi-2}{2\pi}\right)$$

$$\frac{m}{d\pi} = \frac{c^2}{G}\frac{3}{2}\left(\frac{\pi-2}{2\pi}\right)$$

$$\frac{m}{d} = \frac{c^2}{G}\frac{3\pi}{2}\left(\frac{\pi-2}{2\pi}\right)$$

$$\frac{m}{d} = \frac{c^2}{G}\,0.854$$

Thus $\frac{m}{d}$ only has to exceed about 85% of $\frac{c^2}{G}$ to achieve hyperspherical geometry and topology, and in the H6D model of the universe, $\frac{m}{d}$ equals 100% of $\frac{c^2}{G}$ if we equate L, antipode distance, with d, diameter.

Thus it seems unlikely that spacetime singularities can feature in the universe, either as an initial condition or as the result of gravitational collapse, because hyperspheres will form instead.

Appendix iii shows that hyperspheres naturally vorticitate, thus preventing further collapse and creating three-dimensional time.

Appendix III The Hyperspherical Universe

Key to symbols.

G = Gravitational Constant.

M = Mass of Universe

m = Mass

c = Lightspeed

d = Density (Mass divided by volume)

A = Anderson Acceleration

a = Acceleration

V_o = Orbital Velocity

r_3 or r, = Three radius of a sphere

r_4 = Four radius of a hypersphere

W = Angular velocity in radians per second

L = Antipode distance in a hypersphere, (L = πr_4)

l = length

I have a hunch that the universe runs on fairly simple algebra/geometry like 'force equals mass times acceleration', or 'energy equals mass times lightspeed squared'.

I suspect that really complex formulae do not apply to fundamental phenomena.

Part 1. The Vorticitating Hypersphere.

'Matter everywhere rotates relative to the compass of inertia with the angular velocity, (W), of twice the square root of pi times the gravitational constant times density'

-Kurt Gödel.

$$W = 2\sqrt{\pi G d} \quad \text{(Equation 1)}$$

(Gödel derived this as a possible solution to Einstein's field equations).

Now substituting the mass of the universe M, and volume of a sphere, $4/3 \pi r_3^3$ for density, and then substituting $3GM/r_3 = c^2$ (the formula for a photon sphere) into equation 1, and then simplifying, we obtain:

$$W = c/r_3 \quad \text{(Equation 2)}$$

A Photon sphere consists of an object about which light approaching it tangentially would go into orbit. Equation 2 shows that the Gödel universe would have an orbital velocity of c, lightspeed, at its circumference, and a centrifugal acceleration of: -

c^2/r_3. This balances a similar centripetal (gravitational) acceleration.

To give a hypersphere the properties of an orbital velocity of lightspeed means that

$W = c/r_4$

So working backwards and inserting the mass of the universe M, and hyperspherical 3-surface volume, $2L^3/\pi$, for density, and $V_o = c = \sqrt{GM/L}$ (the formula for a hypersphere with an orbital velocity of lightspeed), we recover: -

$W = \sqrt{2\pi G d}$ (Equation 3)

This shows the vorticitation of a hypersphere, in which the entire 3 dimensional surface rotates relative to the orthogonal curvature axis.

Such a structure has a centrifugal acceleration of: -

$A = c^2/L$, (Equation 4)

Part 2. The Size of the Universe.

A universe consisting of a hypersphere with $V_o = c$, has the equation; -

$GM/L = c^2$ (Equation 5)

And thus a centripetal (gravitational) acceleration of, $-a = c^2/L$ to balance the centrifugal acceleration in equation 4.

Now if we equate the Anderson acceleration A,

(Measured at 8.74×10^{-10} metres/second^2), with the centripetal/centrifugal accelerations in a vorticitating hyperspherical universe, then we can easily calculate L and M, and also the temporal horizon of the universe T, to yield the following values: -

$M = 1.39 \times 10^{53}$ kilograms.

$L = 1.03 \times 10^{26}$ metres, about 11 billion light years.

$T = 3.34 \times 10^{17}$ seconds, about 11 billion years.

Angular rotation = 0.006 arc-seconds per century.

Note that these figures have an uncertainty of about 15% arising from difficulties in precisely measuring the Anderson acceleration. The universe will actually look a little larger than L and T because of hyperspherical lensing.

Part 3. The Anderson Acceleration.

The centripetal/centrifugal effect of the Anderson acceleration in a vorticitating hypersphere gives rise to an omni-directional resistance to linear motion and an omni-directional boost to any kind of gravitational orbital motion.

As $a = c^2/L$, light from antipode distance becomes redshifted to oblivion creating effectively an optical horizon.

$C - AT = 0$ (Equation 6)

The Anderson acceleration boosts orbital velocities according to the following equation:

$$V_0 = \sqrt{Gm/r + rA} \quad \text{(Equation 7)}$$

This makes negligible differences at planetary distances, but at galactic distances it makes significant differences, and it obviates the need for arbitrarily modified gravity theories or dark matter.

Part 4. Closed Time Curves.

Gödel's rotating universe idea became discarded as unphysical for two reasons. Firstly no axis of rotation seemed observable. However in a hypersphere the r_4 axis lies at right angles to 3d space and remains unobservable except as curvature.

Secondly the Gödel universe contains closed time curves and anything travelling around the universe at lightspeed would in theory eventually catch up with its own past, in the sense that it would arrive back just as it began to set off.

In the vorticitating hyperspherical universe exactly this happens, but it does not create a causality problem, rather it solves the problem of causality by making everything the cause of everything. However no form of radiation or matter could in practise survive the 22 billion year trip and expect to arrive in the same form it departed in.

Part 5. Hyperspherical Particles.

Equation 3, for the angular velocity of a hypersphere,

$$W = \sqrt{2\pi Gd}$$

contains a further surprise.

It reduces to $W = c/r_4$, and substituting $W = 2\pi f$, to find the frequency f, and then substituting $(L = \pi r_4)$ yields:

$$f = c/2L \quad \text{(Equation 8)}$$

Now if we identify L with wavelength then this equation also represents the basic unit of fermion particle spin, where one half of frequency times wavelength equals lightspeed. This also explains why fermions have to rotate through 720 rather than 360 degrees to restore their original orientation.

Thus it seems that fundamental particles consist of vorticitating hyperspheres as well. This seems inevitable if they have the rotational freedom described by HD8.

Thus Equation 3 unites the Microcosm and the Macrocosm.

I suspect that Hermes Trismegistus would have appreciated that.

I suspect that the Sufis would also appreciate confirmation that everything spins, including the universe itself.

figure 1 The Hypersphere Projection

figure 2 Geodesics in the Hypersphere Projection

figure 3 'Flattened' Hypersphere

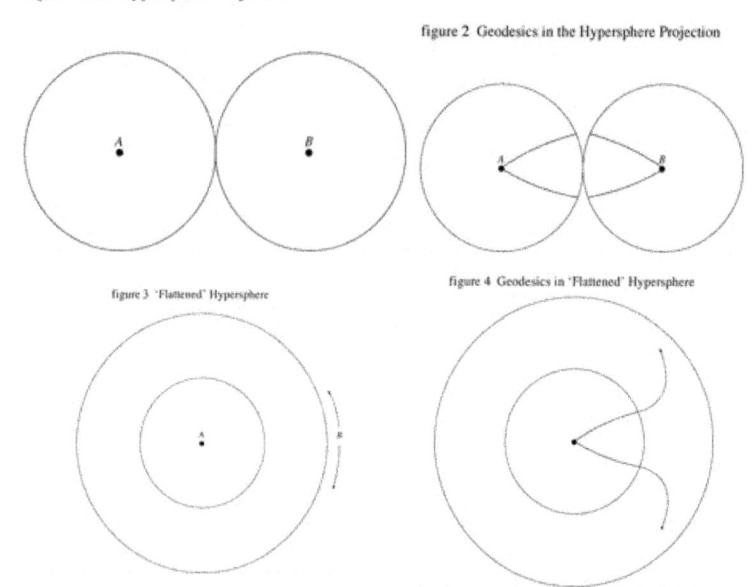

figure 4 Geodesics in 'Flattened' Hypersphere

figure 5 Hyperspherical Lensing

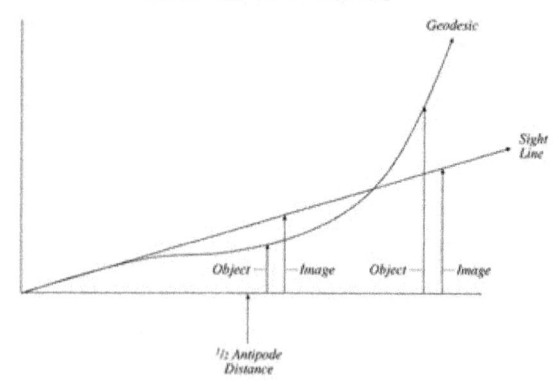

Appendix IV The Shape of the Universe

If you live in a hyperspherical universe with a positive space-time curvature but you assume that you live in a flat universe instead, then you will run into strange problems. You will basically end up with worse versions of the problems of horizons and edges that arise if you persist in believing in a flat earth.

A non-infinite universe must have a definite shape and size, but the finite and unbounded hypersphere or 3-sphere which the universe probably consists of does not easily submit to visualisation unless we remove one of the spatial dimensions for illustrative purposes.

The polar type projection mentioned in chapter 6 results from cutting the hypersphere into two hemi-hyperspheres which we can represent as spheres shown by circles in Figure 1.

These two circles represent spheres whose perimeters contact each other at every point on their surfaces. We can imagine this by allowing the spheres to roll freely around each other.

Position A represents an observer in a hypersphere where we have chosen slice it into two hemi-hyperspheres to position the observer in the centre of one of them. We could have cut it anywhere for illustrative purposes, a hypersphere contains no special positions in reality.

Now an observer at position A can set off in any direction and eventually reach position B, an antipode point which represents the furthest distance you can travel from A without starting to return towards it. All straight-line routes from A lead to B, in much the same way that all straight line trips from the North Pole of the earth lead to the South Pole. See figure 2.

In a hypersphere a straight line route, the shortest distance between two points in 3-dimensional space, has to follow the gravitationally induced curvature of the universe itself. Light also has to follow such routes, which we call geodesics.

Now we always construct an image as though light had travelled to us in a straight line. A lens or mirror actually bends the path of light, but because we construct images on the basis of the direction in which light approaches us, objects appear magnified or diminished by lenses or repositioned by mirrors.

When we look out into the cosmos we assume that light has come towards us in straight lines and that the apparent position of objects represents their actual positions.

This works reasonably well for short distances but at cosmic distances the curvature of space-time itself acts like a gigantic lens.

If we assume flat un-curved space then we can represent that by un-rolling the whole of one of the hemi-hyperspheres around the other. See figure 3. Here the antipode point of an observer at A has become spread out right round the horizon. This corresponds to the South Pole of the earth lying in every possible direction from the North Pole. If this planet had such an enormous density that it bent the paths of light around its surface, we would see something like this.

Figure 4 shows what happens to lines of sight in a hypersphere, they curve inwards towards the halfway to antipode distance, and then diverge towards the antipode, from the perspective of an observer who assumes flat space.

Thus, as Figure 5 shows, objects around the halfway to antipode distance will appear magnified whilst objects further away than that will appear diminished, because observers assume that they see in straight lines in un-curved space.

Now light travelling down those geodesics towards an observer will become redshifted to lower energies, and if the observer assumes a flat spacetime, this redshift will become interpreted as an expansion of the universe. However because hypespherical spacetime acts as a giant lens, the observer will notice a mismatch between the apparent magnitudes of objects at various distances and their apparent recession velocities calculated from redshift. High redshift objects will appear fainter, and thus apparently further away than they ought to. Thus our befuddled observer may conclude that not only does the universe expand, but that its expansion rate has speeded up during the expansion.

Of course neither of these things has actually occurred. It just looks like that because we inhabit a finite and unbounded universe of constant size whose curvature distorts what we can see.

Appendix V
Apophenia's Birthday

Part 1. Theosynthesis and syncretism.

A Mage may gift the world with the naming of a god.

In simpler times, in aeons past, Mages realised god forms corresponding to the basic impulses of love and war, sex and death, fear of the wild-wood, desire for wealth and power, and so on. They also realised other gods to encapsulate the 'souls' of cities or tribes and the lesser functions of the main gods.

In these more complex times we have need of other gods as well, to complete the occult pantheon.

Deo Duce, Samuel MacGregor Mathers, gave us the Holy Guardian Angel concept, which he presented in his Grimoire, based on fragments attributed to Abramelin the Mage. It represents his final understanding of the Higher-Self, Secret Chiefs principle.

Therion, Aleister Crowley, gave us the Aiwass-Horus-True-Will god form in his Book of the Law. It represents his final understanding of The Beast Within; in all its glory and horror, which lurks beneath the veneer of civilisation.

Zos, Austin Spare, gave us Kia, or at least many think he did, for he wrote in riddles.

It appears to resume his final understanding of the panpsychism underlying all phenomena and represents the basic omnivorous chaotic 'life force' beneath the self-image.

Stokastikos, the author, now offers Apophenia, a goddess to embody the occult style of thought itself, which seeks out the hidden connections between seemingly unconnected phenomena, and the strange meanings and inspirations that these connections may bring.

Each of these gods appears to have stalked its priests throughout life as a sort of shadow-like genie until it finally identified itself and phenomenised.

The author developed an involuntarily hyperactive imagination from an early age, and whiled away his schooldays taking fountain pens on epic interstellar voyages, whilst broadcasting telepathic reports of earth news for the benefit of any passing aliens. Pencil sharpeners became models for vast temples to nameless gods in Amazonian jungles.

Academic performance rarely rose above the avoidance of punishment level.

In adolescence a marked tendency for contrarian and ornery thought developed. Everything seemed questionable and dubious except the exercise of thought and imagination itself. Forbidden, discarded, disgraced, and speculative ideas became particularly attractive and fascinating.

In adulthood, daydreaming became a full time occupation with work breaks fitted in as an afterthought. Well one has to eat and provide for others.

The author still disdains to drive powered vehicles; the scenery just seems to set off too many tangential lines of thought and too many mild hallucinations for safety at speed.

Yes, I have always had Apophenia, or rather She has had me. I didn't even know she had a name until the word came to my attention——

So why do I commend Apophenia to my magician readership?

(I can hardly imagine a civilian having read this far.)

Well, since the Eighteenth Century Enlightenment the human enterprise has become increasingly ideas driven. The sum total of ideas increases exponentially nowadays, doubling perhaps every five years or so, and nobody can hope to keep track of all of it. The problems that humanity faces from the application of many of these ideas also seem to increase exponentially, but we cannot turn back now, we need even more ideas to solve the problems that our ideas have already created.

However we do need ideas of a different kind. We need to develop a more holistic view of how phenomena connect with each other. We need to develop an ecology of ideas to see how they fit together, otherwise our lives will become a cacophony of disconnected and largely meaningless experiences whose pursuit will wreck this planet's environment.

We have experts and specialists for everything now, but knowledge as a whole becomes steadily more fragmented and our identities and societies follow suit.

We need Synthesists, Polymaths, Inter-disciplinarians, and Visionaries, we need Apopheniacs of all varieties - her kind of people.

Klaus Conrad coined the word Apophenia in 1958, and defined it as the "unmotivated seeing of connections" accompanied by a "specific experience of an abnormal meaningfulness".

However the full extent of the 'abnormal meaningfulness' did not become apparent until the spring of 2005.

Apophenia phenomenised as a deity in cyberspace on May 26, 2005 at 11.59.46 AM. No previously manifesting deity has had such an accurately recorded time of birth, although Mithras appeared on December 25th, circum 2nd century BC, but some other religion eventually hijacked that date.

The nativity occurred during a transatlantic exchange between the magicians Corlian and Stokastikos and the priestess Dalryada, on the Maybelogic Academy. It happened during an Augoides discussion in the first Internet course on Chaos Magic.

(The author particularly wishes to also thank the very excellent online Maybelogic Academy itself that made this event possible, and Ratatosk the Squirrel of Discord, the technomancer who recovered the data.)

The text of the exchange follows: -

Part 2. A Surprise Birthday

Corlian, May 26, 2005, 18:25:42 GMT
"Concerning insanity, and I intend this question for everyone — Does apophenia or pareidolia exist? Do they play the role of myth or threat? Even if patterns lie within everything, can we make a mistake by misdiagnosing a pattern, or does the pattern imply/require that initial misdiagnosis?"
                    Stokastikos, May 26, 2005, 18:59:46 GMT

(Having spent the last 20 minutes looking up the word Apophenia on Wikipedia, and undergoing a major epiphany)

"Apophenia, Apophenia, APOPHENIA!
A concept in a word,
A word sonorous enough to name a Goddess,
Yes I most certainly feel we need her in a Chaoist pantheon.
Thank you, Pete."
                    Dalryada, May 26, 2005, 19:23:41 GMT

"— Personally, I'm just thrilled that we've all — scant minutes ago — witnessed the birth of a GODDESS."

May 26 makes Apophenia a Gemini. Subsequent discussions amongst the visionaries of Arcanorium College reveal that Pareidolia must share the same birthday, as her twin. As more of a magician and scientist than a priest or artist I naturally earthed the Apophenic aspect of the twin current. I concluded that whilst Apophenia could bring the Universe in a grain of sand to our attention, Pareidolia merely distracts us with the face of the Virgin Mary in a pavement pizza. Chaos theologists then pointed out the critical importance of Pareidolia in art and mystical religion.

Salvador Dali and whosoever wrote the Book of Revelations must rank as high priests of Pareidolia.

I leave it to the magician-artists and neopantheist-mystics to reveal her formal invocations.

Epilogue

The example Chaos Magic Paradigm presented in the preceding chapters represents a distillation of ideas and evidence available at the time of writing and summates the last decade of my researches.

The evidence suggests to me that we inhabit a quantum-panpsychic universe consisting of a vorticitating hypersphere, that has finite and unbounded extent in both space and time, and that has equal spatial and temporal dimensionality.

The hyperspherical vorticitation of the universe leads to three dimensional time and this provides an explanation for both the strange behaviour of the underlying quantum realm and the occasional appearance of seemingly magical effects in the macroscopic world.

The statistical effects of random quantum behaviour create a semblance of causality in the macroscopic realm which disguises the underlying chaos but 'Magic' basically structures the universe and keeps it functioning, we participate inadvertently and panpsychically in this process.

Yet we can participate directly by deliberate acts of magic, and it works often enough to justify the effort.

The geometry of the vorticitating hypersphere permits magic and invites us to become apprentice gods.

We have worlds within us. Beneath the veneer of the everyday self we have multiple minds. Thus the Neopantheist style of Mythos belief better reflect our psyche than the Logos style of belief.

The retroactive effects of panpsychism may well explain the general features of this participatory universe.

However I have apophenia, and others may see things differently, we seem to have a lot of alternative realities kicking around these days.

Lastly, some Chaoists may feel uneasy with the idea of a six-dimensional universe. I should perhaps point out that the hidden curvature dimensions of space and time, do in a way constitute another 2 dimensions, each having one pi-th the size of the observable ones.

So that makes a reassuring eight.

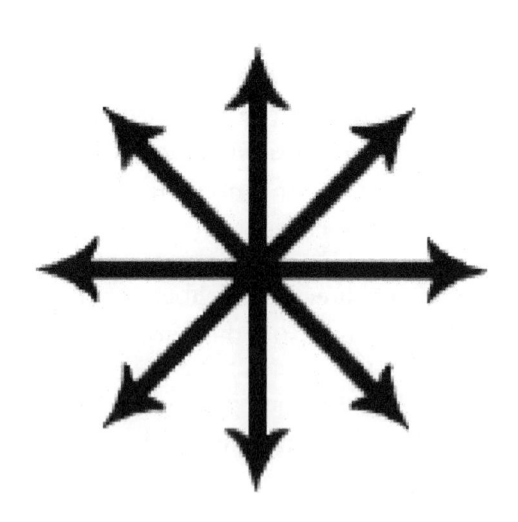

Notes, References, and Bibliography

0) Stochastic, means relating to, or characterised by conjecture and randomness. In a stochastic process, non-deterministic behaviour means that a state does not fully determine its next state. Etymology: Greek stochastikos skilful in aiming, from stochazesthai to aim at, guess at, from stochos target.

1) Illuminates of Thanateros. www.iot.org.uk/pages chaosmagic.html

2) Arcanorium College www.arcanoriumcollege.com

3) Alfred North Whitehead. Philosopher and Metaphysician. 1861-1947.

4) David Chalmers. *The Conscious Mind: In Search of a Fundamental Theory* (1996). Oxford University Press.

5) Nicholas Humphrey in *What We Believe by Cannot Prove.*, Edited by John Brockman. The Free Press.

6) Chalmers, see 4.

7) *The User Illusion: Cutting Consciousness Down to Size.* by Tor Norretranders Penguin 1999

8) *Multimind: A New Way of Looking at Human Behavior* by Robert E. Ornstein.   Publisher: Ishk 2003

9) Adapted from *S.S.O.T.B.M.E. Sex Secrets Of The Black Magicians Exposed, An essay on magic.* By Ramsey Dukes. Pub, The Mouse That Spins.

10) *Liber Kaos*, by Peter J Carroll. Pub, Weiser 1992. See Aeonics chapter.

11) Penrose - *The Road to Reality: A Complete Guide to the Laws of the Universe*. Vintage Books 2006

12) The Copenhagen Interpretation of Bohr and Heisenberg. Circum 1927. Physics concerns what we can say about nature, it cannot tell us what it actually 'is'.

13) *Hidden Variable Interpretation*. De Broglie ~ 1927, extended by Bohm ~ 1952. This preserves determinism but locality fails, information travels faster than light.

14) *An Overview of the Transactional Interpretation* by John Cramer. International Journal of Theoretical Physics 27, 227 (1988) A Farewell to Copenhagen?, by John Cramer. Analog, December 2005.

15) *A Brief History of Time: From the Big Bang to Black Holes* By Stephen W. Hawking. Published 1988.Bantam

16) Liber Null & Psychonaut by Peter J Carroll. Pub, Weiser 1987. See Gnosis section.

17) See10. Part 2, *The Psychonomicon*, sleight of mind.

18) Waldo Thompson. www.http://hdcity.com/cosmos

19) *The Trouble With Physics: The Rise of String Theory, The Fall of a Science, and What Comes Next* by Lee Smolin. Mariner Books 2007

20) *Not Even Wrong: The Failure of String Theory and the Search for Unity in Physical Law* by Peter Woit. Mariner Books 2007

21) *Principia Discordia* - Malaclypse the Younger

Further reading.
Authors Website.www.specularium.org

Uncle Ramsey's little Book of Demons. By Ramsey Dukes. Aeon Books 2005.

The End.

Index

A
Anderson acceleration 96, 98, 141
Anusites 47
Astronomicon 118, 119, 120, 121

B
big bang theory 88, 90, 92, 99

C
Cardano 22
Chalmers 22
Chaoism 7, 9
Chaoist 41, 52
Conrad, Klaus 150
Council of Nicea 49
Crowley, Aleister 148

D
Darwin, Charles 20
Descartes 29
Double Slit Experiment 66, 71, 75

E
Eris 113, 117, 118, 119, 121

F
free will 23, 24, 25, 27, 37

G
Giordano Bruno 22
Gnosis 78
Gödel 95
Incompleteness Theorem 99

H
Hawking, Stephen 106

L
Leibniz, Gottfried Wilhelm 22, 115

M
Mathers, Samuel MacGregor 148
mind, theory of 35, 54

N
Noether's theorem 73
Norretranders 36

O
Ornstein 36
Ouranos 9, 117

P
paganism 37, 50
Pareidolia 8, 9, 114, 117, 118, 119, 121, 151
Penrose, Sir Roger 64
Philosophical Zombie 31, 34
Plato 48
psychogram 52

Q

Quantum physics 7, 22, 64, 65, 66, 70, 75, 86

R

Ramon Lull 115

S

schizophrenia 38, 39
Schopenhauer 22
Sky Fairies 52, 60, 61
Spare, Austin 148
Spinoza 22

T

Thales 22

W

Whitehead 22

Mandrake

'Books you don't see everyday'

Bright From the Well by Dave Lee
978-1869928-841, £10.99

'Bright From the Well' consists of five stories plus five essays and a rune-poem. The stories revolve around themes from Norse myth - the marriage of Frey and Gerd, the story of how Gullveig-Heidh reveals her powers to the gods, a modern take on the social-origins myth Rig's Tale, Loki attending a pagan pub moot and the Ragnarok seen through the eyes of an ancient shaman.

The essays include examination of the Norse creation or origins story, of the magician in or against the world and a chaoist's magical experiences looked at from the standpoint of Northern magic.'

'Dave Lee coaches breathwork, writes fiction and non-fiction, blends incenses and oils, creates music and collages'

Magick Works: Pleasure, Freedom and Power by Julian Vayne
978-1869928-469

Enter the world of the occultist: where the spirits of the dead dwell amongst us, where the politics of ecstasy are played out, and where magick spills into every aspect of life.

It's all right here; sex, drugs, witchcraft and gardening. From academic papers, through to first person accounts of high-octaine rituals. In Magick Works you will find cutting edge essays from the path of Pleasure, Freedom and Power.

In this seminal collection Julian Vayne explores;

* The Tantric use of Ketamine.
* Social Justice, Green Politics and Druidry.
* English Witchcraft and Macumba
* The Magickal use of Space.
* Cognitive Liberty and the Occult.
* Psychogeography & Chaos Magick.
* Tai Chi and Apocalyptic Paranoia.
* Self-identity, Extropianism and the Abyss.
* Parenthood as Spiritual Practice.
* Aleister Crowley as Shaman

...and much more!

Other Mandrake Titles:
Fries/*Cauldron of the Gods: a manual of Celtic Magick*.
552pp, royal octavo, 9781869928612 £24.99$40 paper

Fries/*Seidways Shaking, Swaying and Serpent Mysteries*. 350pp
9781869928360 £15/$25
Still the definitive and much sought after study of magical trance and possession techniques.

Fries/*Helrunar - a manual of rune magick*. 454pp
9781968828902 pbk, £19.99/$40 Over 130 illustrations. new enlarged and improved edition
'...*eminently practical and certainly breaks new ground.*'
- Ronald Hutton

Order direct from
Mandrake of Oxford
PO Box 250, Oxford, OX1 1AP (UK)
Phone: 01865 243671
(for credit card sales)
Prices include economy postage
Visit our web site www.mandrake.uk.net

Lightning Source UK Ltd.
Milton Keynes UK
UKHW041227080720
366216UK00001B/70